THE YEAR OF THE ANT

BY THE SAME AUTHOR

Untaken Harvest

Wine Growing in England

Garden Pests

The Living House

The Last of the Incas, with Edward Hyams

Man, Crops and Pests in Central America

Pigeons and People, with Pearl Binder

Biological Methods in Crop Pest Control

Ladies Only, with Pearl Binder

The Great Wine Blight

John Curtis and the Pioneering of Pest Control

The Year of the Butterfly

The Constant Pest

THE YEAR OF THE ANT

George Ordish

ILLUSTRATED BY

Clarke Hutton

 CHARLES SCRIBNER'S SONS / NEW YORK

Library of Congress Cataloging in Publication Data

Ordish, George.
The year of the ant.

Bibliography.
Includes index.
1. Formica rufa. 2. Ants. I. Title.
QL568.F707 595.7'96 78-2260
ISBN 0-684-15523-0

THIS BOOK PUBLISHED SIMULTANEOUSLY IN THE UNITED STATES OF AMERICA AND IN CANADA— COPYRIGHT UNDER THE BERNE CONVENTION

1 3 5 7 9 11 13 15 17 19 V/C 20 18 16 14 12 10 8 6 4 2

PRINTED IN THE UNITED STATES OF AMERICA

To Olive

CONTENTS

PREFACE

I am much indebted to the writings of many authors, living and dead, for material used in this book, and I thank them. I also thank a number of librarians for their help: The British Library, London; Harvard University; The New York Public Library; Rothamsted Research Station, Harpenden, England; The Royal Entomological Society, London; and The Tropical Products Institute, London.

Many individuals have helped me too, in particular Gary Alpert, John T. Bonner, Stephen Edwards, Jack and Linda Grobstein, Peter Walker, Michael Way, Edward O. Wilson, and many others. Finally, I thank my wife for much help in constructing this book and in the revision of the text.

THE YEAR OF THE ANT

1

THE ANT WORLD

ANT workers are all females, but not fully developed ones. The ant world has a profound and proper respect for males. They do no work at all and lead a short but glorious life with one aim in view: the propagation of the species. Male ants are fortunate in that they have been able to preserve their sexual niche over the eons of the species' existence. Had wood ants, the subject of this story, found a way of doing without males, they would have dispensed with the need to spend scarce resources (food and labor) on raising large numbers of males every year. Such wasted materials could then have been used to increase leisure among the workers (ants enjoy leisure; sunbathing, for instance) or to extend the range and power of the species. Ants generally might then have dominated the earth instead of man.

But this raises the point whether ants do not already dominate the earth. According to Edward O. Wilson, a leading living authority on the ant, seventeenth-century Portuguese settlers in Brazil called the ant the king of Brazil and maintained that the country was one great ants' nest. Ants are found everywhere, from desert to tropical rain forest, from the equator to the edges of the Arctic and Antarctic circles; they are far more numerous than man. They achieve everything they attempt to do; they take advantage of new environments (houses, for instance) and colonize them. Although ants, over the whole world, as individuals are far more numerous than man, their biomass is less, which can be an advantage in the survival stakes, because they do not have to feed and support so much animal tissue. Even so, E. O. Wilson wrote that "their biomass and energy consumption exceed those of vertebrates in most terrestrial habitats." So they are running man a close second in the race for domination, if they have not already won it. Even when their interests clash with man's and he attacks them, man is not necessarily victorious. The Argentine ant (*Iridomyrmex humilis*) continues to plague fruit trees the world over in spite of control measures by man, except in Argentina—because there it is called the Brazilian ant and does just as much damage. In the United States the fire ant's attacks got worse after thousands of acres had been sprayed with powerful insecticides. Maybe the ant is the world's dominant animal without knowing it and just takes man in its stride.

In the popular mind, which relies largely on King Solomon,[1] ants are very hard-working, industrious, and effective, but in fact they are not very efficient if we compare results with effort expended. They learn very slowly and change their habits yet more slowly, frequently failing to adopt a course of action staring them in the face that would save the colony immense trouble. For instance, in the summer one stream of ants may be seen, each ant carrying something such as grass seeds into a nest, and another stream on the other side of the nest may be seen carrying the same seeds out again—wasted labor.

Ants have achieved high success despite these drawbacks; they are found everywhere, except in the polar regions. One of the main reasons, first pointed out in 1925 by the great American myrmecologist William Morton Wheeler, has been their economy in the use of nest-building materials. They mostly use earth, an existing material, for their structures and need not elaborate expensive substances such as wax, as bees do, or paper, as do wasps.

Ants belong to the insect order of Hymenoptera, which includes wasps, bees, sawflies, and certain parasitic insects such as the ichneumons. The Hymenoptera have two pairs of membranous wings, the hind pair being smaller. The wings are interlocked with a pair of hooks (hence *hymen*, or joining together) so that they beat as one during flight. Among ants only the queens and the males now have wings. After mating, the incipient queens shed their wings and the males are either killed in the act or die shortly afterward, their utility being over.

Ants have a very long history, as may be seen from their fossils. Ants were quite numerous in the early Tertiary period, about 50 million years ago. These Tertiary ants show the modern differentiated forms—males, females, and workers—suggesting that they descend from a much earlier period, probably the Triassic, about 175 million years ago; but no remains of those earlier ants have been found. Ants are small, delicate creatures that do not leave much of a record, except when trapped in amber (see page 6).

The great Swiss myrmecologist Auguste Forel thought the precursors of ants were the asocial mutillids, another Hymenopterous insect. Today the Mutillidae are widespread, asocial, parasitic, wasplike insects living on the larvae of solitary bees, wasps, and some flies. The pre-ants (the mutillids) must gradually have developed social habits, the change in their psychology being as important as any physical alteration in their structure.

Today the generally accepted theory explaining changes in living forms (for example, animals, plants), leading to the creation of new species, is the Darwinian struggle for existence. It was first set out by Charles Darwin in 1860, in his book *Origin of Species by Means of Natural Selection*. The theory, though still held true, has been pushed further back by some modern biologists and seen as a struggle for existence by the very genes themselves. Among living forms, competition for food and living space is fierce. Those combinations of genes, packed into certain living forms ("survival machines," to use Richard Dawkins's expression) and possessing some advantageous quality—physical or mental—over their rivals, are in fact the species that survive. Darwin called it natural selection.

With certain exceptions, particularly in ants (see pages 112–13), offspring are not exactly the same as their parents, since half the genes they carry are derived from the male, half from the female. Thus, the opportunity for selection arises, and the more generations there are, the greater are the opportunities for changes to take place. Advantageous changes under the existing environmental circumstances accumulate as time goes on until a new creation—possibly a new species—is produced.

Early man, it will be recalled, dates from about 10 million years ago, but modern man (*sapiens,* if that is the right word) has only a half-million years of existence. Ants have had much more experience of life. Comparing the number of generations in the history of ant and man, the ants at 7 years and man at 20 per generation, we find that even from Tertiary times (and ants must have existed before then), ants have had 7 million generations to man's 25,000 from the time of his "improved" beginnings. Comparing the opportunities for change, we find that ants seem to have had at least 280 times more chances (7,000,000 ÷ 25,000 = 280) than man but seem scarcely to have used them to change their bodies, numerous as the generations were. For instance, Anton Handlirsch noted that of six hundred Hymenoptera found in Tertiary deposits, half were ants. Much evidence from those remote geological times is found in amber. Amber, from the Eocene (50 million years ago), is the fossilized resin from pine trees, which often trapped ancient ants in it. Suitable solvents will dissolve the resin, and the ants contained in a piece can then be studied.

William M. Wheeler found forty genera of ants in amber. Thirteen were extinct species and twenty-seven (more than two-thirds of them) were specimens of still-living kinds. Ants probably originated in North America and spread to Europe and Asia when the continents were joined together with bridges passing through Greenland, Iceland, the Faroes, Spitzbergen, the Aleutian Islands, and Alaska. Ants, like sharks, seem to have hit on the right body formula early in their career and have found very little need to change it since.

One of the difficulties in appreciating and understanding ants is anthropomorphism. We tend to look on them in human terms;

Solomon, already mentioned, is one of the earliest examples of this attitude. But ants are very nonhuman. It is possible to compare human and ant behavior not as anthropomorphism but as an interesting tracking of similarities and differences, for much ant behavior, seemingly similar to human reactions, is effected by nonhuman means. For example, ants secure much of their food from greenflies and scale insects on trees. If the heavy traffic of ants going up and down a tree so infested is stopped by putting a sticky band around the trunk, the ants cannot reach their "cattle" and become desperate, because they have lost a territory and a valuable source of food. Some apparently "noble" ants, usually among those trying to get down the tree, push onto the band, become entangled in the sticky substance, and die there, worthy recipients of the Myrmecan Purple Heart, one might say. A bridgehead is thus established and soon more and more ants, treading on their dead and dying companions, make a bridge of dead ants and normal traffic is resumed, thanks to the sacrifice of a few hundred workers.

In fact, of course, no such sentiments inspire the ants. The bridge is made because the stream of ants, interrupted by the sticky band, circles around and around the tree looking for a way out. Sooner or later the pressure of ants (usually from above) pushes an ant onto the band. The creature gets stuck and puts out an aromatic alarm substance and one or two more ants may go to help it, entangling themselves. The alarm substance increases the speed and numbers of the circling ants and, unable to be as careful in their movements, more and more of them get stuck. Soon the bridge is completed and regular traffic is reestablished. Natural selection so far has enabled ants to solve this problem in a certain way, but sacrificing several hundred ants to pass over each band is not necessarily the best solution in terms of gene survival. It is, however, a hazard of but recent date to ants—of about seventy years' standing—and, moreover, such bands are not often used now, so that a gene combination allowing ants to develop a better method of overcoming sticky bands is not likely to arise in the Argentine ant, the main pest against which sticky bands were used. On the other hand, there are some ants that use little lumps of earth, straws, and twigs, as well as their own bodies, to cross such bands.

Ants are part of a highly competitive system of living forms seeking survival and expansion. In regard to survival they have succeeded remarkably well so far. In fact, if man destroys himself, as he appears determined to do with atomic warfare and radiation, ants have a fair chance of succeeding us as the dominant species on earth, particularly because, according to a modern entomologist, Rémy Chauvin (1970), ants can withstand large doses of radioactivity. It behooves us to understand our successors and, maybe, give them a few useful tips, though it is difficult to see how that could be done.

An example of our anthropomorphic thought in regard to ants is the use of the word *queen*. The queen ant, in the accepted sense of the word, is no such thing. She rules nothing and gives no direct commands. Admittedly, she gives some indirect orders, for certain secretions that she gives off influence behavior in the whole nest. The queen in turn is influenced by secretions given off by the larvae. From the entomological point of view the queen ant is just a fully developed female whose task is to lay fertilized or unfertilized eggs, according to the needs of the community, throughout the three to fifteen years of her life.

In this book we must try to avoid the anthropomorphic view and take, as far as possible, an ant's attitude to life—a myrmecan one. A difficulty is that even saying "an ant's attitude" is being anthropomorphic. Ants do not have an "attitude to life." Their scale of awareness is limited, yet they are not just mechanical, just part of the machine. There is a modicum of free will in their system, which is shown by at least three things: ants enjoy leisure, individual ants are capable of learning (the shortest way to food, for example), and they make choices. But the scale of such activities is very limited and, when indulged in, can be overridden at any moment by the needs of the machine. It must be kept running at all costs. The control exercised is largely chemical: secretions emanating in the nest start and stop activities needed by the colony.

Ants have the habit of regurgitating food and exchanging it

with other ants, a process called trophallaxis. It is a common sight to see one ant meet another and touch antennae, after which one of them will bring up a droplet of liquid into her mouth and offer it to the other creature, which then absorbs it. If a drop of blue-dyed food is given to a few ants it can be noted that within a few hours the stomachs of nearly every ant in the nest will be blue. Chemical messenger substances in the food—pheromones—are thus quickly disseminated throughout the colony. There is a human parallel to this: we exchange air in much the same way. For instance, at a cocktail party every time you smell tobacco smoke you are taking in air that has just been in someone else's lungs. As far as we know, the reabsorbed air contains no chemical controlling human behavior, except perhaps one strengthening addiction to tobacco.

Ants have a collective stomach in much the same way as they may have a collective brain (see page 18). If the nest is short of food, the "go foraging" signal is given. If the queen dies, the "no queen present" message flashes around the colony and appropriate action is taken. The actions are automatic and arise from the needs of the colony as a whole, not from those of any individual. The social system of the ants might be described as the only known example of the Marxist system in successful practice. Each ant gives according to its ability and receives according to its needs. The state has withered away, for, as Solomon pointed out, ants have "no guide, overseer or ruler." Chemistry controls the community.

Ants are typical insects. An ant's body is divided into three sections—head, thorax, and abdomen—and each ant has six legs, attached to the thorax. The largest of the wood ant workers are of moderate to large size, about one-third inch long. They are a light or dark brick red in color, with the end of the abdomen black or brownish. Wood ants were given the scientific name *Formica rufa* by the great naturalist Carl von Linné (whose name is usually Latinized to Linnaeus) in 1758. Today so many closely related *Formica* genus ants have been described by scientists that the naming of species has become complex and a source of dispute among experts. Naturalists now refer to the *Formica rufa* group. As early as 1905 William M. Wheeler had recorded two subspecies. By 1951 eighteen were noticed by C. F. W. Muesebeck and his coauthors, and the number continues to grow, like the ants.

Ants of any given species usually exist in three main forms, as workers, males, and females, the latter usually being called queens. Workers, the most numerous clan, are underdeveloped females. In writing about them they are usually given the astrological sign of Mercury ☿, distinguishing them from Venus ♀ (queens) and Mars ♂ (males). In any one species worker ants of different sizes may be found, though among wood ants the differences are not as great as those found in some kinds. Small workers one-fifth inch long are sometimes seen.

As do all insects, ants lack a backbone. Their hard shell, or the exoskeleton, gives support to all their muscles and organs. Once hatched from its pupa, the adult ant grows no more. Wood ants of different sizes were born those sizes. The Formicoidea have a complete metamorphosis; they pass through the stages of egg, larva, pupa, and adult.

Wood ants have elaborate mouths with a stout pair of mandibles, each bearing eight teeth. These mandibles are their universal tools for attack, defense, transporting goods, digging, building, and cutting up food. They can be worked independently from the rest of the mouth. The mouths also include a pair of maxillae, or jaws, with various organs on them, used to handle, push in the food, and get it into edible form. The maxillae bear two sensing palps, which help them pass judgment on the suitability of food and interpret the chemical messages presented in it. The labium, or lip, in the mouth acts as a tongue and is also a complex structure. It is ridged so as to rasp food, breaking it up into small particles that can then be swallowed. When not in action the labium closes the mouth. The ant uses its tongue to clean itself and its colleagues.

The head carries a pair of antennae, a vital feature of the ant's anatomy and the chief sensing organ. The antenna of workers and queens has twelve sections, that of males an unlucky thirteen—males, as we know, have a very short life. An antenna is joined to the head by a ball-and-socket joint similar to that in the human shoulder, and can therefore be moved through a wide field, smelling and feeling the surrounding air and objects. The first length (the one springing from the head), known as the scape, is comparatively long. The remaining sections are small, though the last is swollen into a tiny club. The antennae bear small hairs by means of which the creature smells. Since the ant is covered with a continuous layer of impervious cuticle, it was once difficult to understand how the creature could smell by means of the hairs. There appeared to be no contact of any nerves with the air. But, as Sir Vincent Wigglesworth has discovered through the modern electron microscope, the hairs bear many minute pores in contact with the atmosphere. Each pore exposes the end of a dendrite, or branched nerve fiber, to the air, making the antennae very sensitive not only to the smells themselves but also to the direction from which they emanate, the two quite mobile antennae being used to locate their origin. The ants also have a "topochemical" sense that enables them to "feel" a smell. By touching something they know what it smells like. Smell and touch are of great importance in ants' underground life, replacing sight. In fact, ants' eyes are used more for hunting than for anything else, though in practice the wood ants can hunt in the open at night almost as well as during the day, relying on smell and touch to do it.

Wood ants have a considerable sense of gravity, knowing when they are climbing up or down a hill or a vertical surface. This sense operates largely by means of the bristles called bristle fields on the joints of the antennae and at the base of the scape. Wood ants also know when they are on a level, which is important in direction finding (see page 38).

Ants cannot exactly hear, but they are sensitive to vibrations carried by the ground or by any surface on which they are standing, through organs situated on their legs.

The head carries five eyes, two that are compound, with many facets, and three that are simple eyes, or ocelli. These are discussed later (see page 44). The thorax carries the six legs and, in the case of the males and queens, the two pairs of wings. The wings are operated by powerful thoracic muscles that become important food reserves after the marriage flight, when the wings are cast off. Workers are wingless.

The thorax is joined to the abdomen by a slender stem known as the scale. The abdomen contains the heart, the digestive and excretory organs, and, in the case of the males and females, the reproductive organs. The rear of the abdomen is the gaster.

The abdomen also contains some important secreting organs, the two most striking being the formic acid gland and Dufour's gland, the latter named for the one who discovered it. Formic acid is both a weapon and a messenger substance for wood ants. It is stored in a poison sac, or vesicle, from which it is conducted by a duct past an accessory gland to the anus, from which it can be ejected as a fine spray or oozed out as a drop. The storage vesicle is large, occupying one fifth the volume of the abdomen. The liquid in it is strongly concentrated—50 percent formic acid. As the acid is insecticidal and corrosive, it is remarkable that it does not kill its owner. The tissues composing the vesicle itself and the ducts must be highly resistant to the chemical. It may be noted that other insects can also produce powerful chemicals, one of the strangest being the bombardier beetle (*Brachinus* spp.). The insect goes *pop* when picked up. It seems to have two glands, one secreting hydrogen peroxide at a high concentration, which is a very active substance, a disinfectant, and an oxydizing agent, and the other making a peroxydase. The mixture of the two substances as it is discharged from the anus produces the tiny explosion and a minute puff of smoke.

Wood ants have no sting, but the fine spray of formic acid from the discharges of thousands of ants is a defense against birds and mammals, such as shrews and man. Another deterrent is the use of acid to back up a direct attack. An ant's powerful jaws can inflict a wound into which the insect injects acid by twisting her abdomen up toward her mouth and squirting the poison into the laceration, causing much irritation. It is all done in a flash and seems to be just like a sting, reminding one of the supposed advantages of the stingless native American bee. Although it has no sting, it can give just as effective a bite. A succession of poisoned ant bites in a nice piece of a human intruder's soft flesh can be most painful and lead to his hasty retreat, as many an investigator has found.

In 1968 F. E. Regnier and E. O. Wilson discovered that the venom of another ant (*Acanthomyops claviger*) is composed of formic acid and a series of aliphatic[2] hydrocarbons and ketones, among them formic acid and undecane, both being particularly volatile. In evaporating, formic acid creates a repellent shield, and undecane, which is both a defensive and a communication substance, broadcasts a continuous alarm signal.

Recently Jan Löfqvist studied formic acid and the Dufour gland secretions in the wood ant. He identified thirty-nine substances in the Dufour output, twelve of them in amounts in excess of 1 percent of the total secretion. Undecane was the most common ingredient (51 percent), followed by tridecane (20 percent), and pentadecane (6 percent). Obviously, all the chemicals were volatile—to be smelled by the ants they would have to be—but their degree of volatility differed.

Löfqvist noted six steps in alarm behavior in ants: 1. walk randomly, stop, raise antennae and wave them inquisitively; 2. open mandibles; 3. walk slowly to odor source; 4. run fast; 5. attack intruder; 6. clean antennae and tip of gaster.

Formic acid proved to be a very strong alarm substance. It quickly produced the number 4, fast-running, reaction and lasted for sixty seconds, when there was a certain concentration (10 billion molecules per cubic millimeter), and then faded. As the evaporation rates were different, a succession of effects was produced, based on the quantity and quality of each substance. The ants probably could measure the amounts of the different pheromones present and react accordingly. Finally, it seemed there was a "stop" signal left, leading, no doubt, to reaction number 6.

An additional accessory gland in the abdomen produces an alkaline substance and can be used to neutralize any acid adhering to the ant's body after the poison has been used.

Ants have no lungs. Air circulates throughout the system by means of tracheae, or fine air tubes, opening to the exterior by means of spiracles, of which there are ten. Spiracles have valves enabling them to be closed if the need arises.

A colorless blood is circulated throughout the ant by means of a simple heart situated in the mid–upper part of the gaster.

Numerous glands exist throughout the body, two already having been mentioned, but there are many others, all exercising some degree of control over the ants and the whole colony. An ants' nest is a gland-controlled entity.

The nervous system consists of nerves joined to ganglia throughout the body, those above the throat (the esophagus) being considered to be the brain.

Ants seem to have fascinated man from biblical times onward and have frequently been used by him to make moral and political points. They have a considerable literature, both scientific and lay. With more than ten thousand different species and many differing life histories, it is obvious that the scientific literature would be considerable. In fact, it might be said that a subspecies of the human race—*Homo sapiens myrmecus*—has arisen to study them, some famous names being Horace Donisthorpe (United Kingdom, d. 1951), Carlo Emery (Italy, 1848–1925), Auguste Forel (Switzerland, 1841–1912), Father Wasmann (the Netherlands, 1859–1931), and William Morton Wheeler (United States, 1865–1937). Among nonscientists, King Solomon has already been mentioned.

La Fontaine's famous fable *Le cigale et le fourmis* (*The Cicada and the Ant*) was meant to point out the virtues of hard work and of taking thought for the morrow, but it in no way conforms to fact. It much annoyed the great French entomologist Jean-Henri Fabre (1823–1915), who pointed out, first, that La Fontaine was wrong in thinking the *cigale* to be a grasshopper, an error projected into the English-speaking world by mistranslation of that word to "grasshopper," and second, that the natural history was all wrong. Quoting a Provençal poem (incomprehensible unless translated from that strange tongue!), Fabre pointed out that in winter the cicada's young were safe in the earth and that she herself died at the end of summer; she fell from her tree and was eaten by the ant ("De tu magro peu dessecado / La marriasco fai becado"). Even if the *cigale* had been a grasshopper, the natural history would still have been wrong. The grasshopper's method of passing the winter as an egg is more economical and just as successful as the ant's hibernation deep in the nest as an adult. It is perhaps some consolation to think that *The Cicada and the Ant* would never have made as good a Walt Disney film as did *The Grasshopper and the Ant.*

Mark Twain, who was anti-ant, to make a humorous point wrote that its virtues had been greatly exaggerated. In 1953 John Compton quoted Twain on what Compton described as a "pearl of witty inaccuracy."

> Science has recently discovered that the ant does not lay up anything for winter use. This will knock him out of literature to some extent. He does no work except when people are looking, and then only when the observer has a green, naturalistic look, and seems to be taking notes. . . . It is strange beyond comprehension that so manifest a humbug as the ant has been able to fool so many nations and keep it up so many ages without being found out.

An eighteenth-century naturalist, George H. Millar, had some surprisingly sound ideas on ants, criticizing, in a much calmer way, the literary treatments that aroused Fabre's anger and Mark Twain's mockery. Millar's accurate observations have rarely been noted in recent times. About 1785 he wrote:

> These insects are famous from all antiquity for their social and industrious habits: they are offered as a pattern of parsimony for the profuse, and of unremitting diligence to the sluggard. It is, however, surprising that all the writers of antiquity should describe this insect as labouring in the summer, and feasting upon the produce during the winter; it being well known that they require no supply of winter provisions, as they are actually in a state of torpidity during that season.

Millar also noted that there were three kinds of ants (males, females, and neuters), that only the two former classes had wings, and that they "mix with the working multitude, but seem no way to partake in the common drudgeries of the state." He also remarked on the ants' devotion to their young: "The fond attachment which the working Ants shew to the rising progeny is amazing: in cold weather they convey them in their mouths to the very depths of their habitation, where they are less subject to the severity of the season. In a fine day they remove them nearer the surface, where their maturity may be assisted by the warm beams of the sun."

These are remarkable observations for a period when there was much ignorance about insect life. For instance, with an important production insect—the bee—many people then thought the queen bee to be the king of the hive and that it ruled there. Presumably they thought that eggs and larvae arose spontaneously from the honey. Spontaneous generation of life was a popular belief until Pasteur killed it in the mid-nineteenth century, though the possibility of such a thing is again being considered by a number of scientists.

Richard Lovelace, a seventeenth-century English royalist, wrote a charming poem on the ant. Unaware that the ant was not the ceaseless worker it appeared to be, the poet urges the creature to enjoy a little leisure:

Forbear thou great good husband, little Ant;
A little respite from thy flood of sweat;
Thou, thine own horse and Cart, under this Plant
Thy spacious tent, fan thy prodigious heat;
Down with thy double load of that one grain;
It is a Granarie for all thy Train.

Cease large example of wise thrift a while
(For thy example is become our law)

. . . .

Austere *and* Cynick: *not one hour t'allow*
To lose with pleasure what thou gotst with pain.

But we must not overdo our concept of the leisure actually enjoyed by ants: it is very little. They can work literally for days on end and whenever the need occurs.

Donisthorpe, the English ant man, was much annoyed that the Swiss myrmecologist Forel (whose scientific work he much admired) should use his book on ants to further the cause of socialism. The English myrmecologist wrote: "An entomological work is not the appropriate means for the introduction of political theories of any kind, still less for their glaring advertisement," and he went on to point out that ant workers were sterile, had no trade union, worked from "morn to dewy eve," and often had their heads cut off when their utility was over.[3]

The Hungarian author Arpad Ferenczy dedicated a novel, *The Ants of Timothy Thümmel,* to Donisthorpe in 1924. It describes how the African paradise of the Aruwimi ants was disrupted (119,000 generations ago!) by Mye-Mye, a male ant, who started to think.

The American playwright Samuel Spewack wrote an amusing play entitled *Under the Sycamore Tree* (1952), about ants, assuming that they had learned to talk and had constructed an amplifying short-wave radio transmitter and receiver. Deciding that ants, too, had a stake in the world and that man might destroy it, the ants try to arrange a conference with the president of the United States.

The last novel Edward Hyams wrote was *Morrow's Ants,* in which by studying ants a captain of industry builds up a gigantic, partly underground factory and colony that is perfectly contented. The behavior of the population is controlled largely by means of pheromones in the air and food. So well organized is it that the assassination of the said captain of industry makes no difference at all, and a successor is appointed. If the queen in a monogynous (one-queen) nest dies, another is raised to replace her.

The fact is, of course, that ants in the "City of Utter Otherness" (Chauvin) are so far removed from our world that they have no lesson for us at all. One studies them because their systems are so strange and because man's social system is not the only solution to the problem of widespread survival imposed by the evolutionary struggle in the world of living things.

Alexander Pope put it very neatly in his *Essay on Man.* Nature, speaking to man, says:

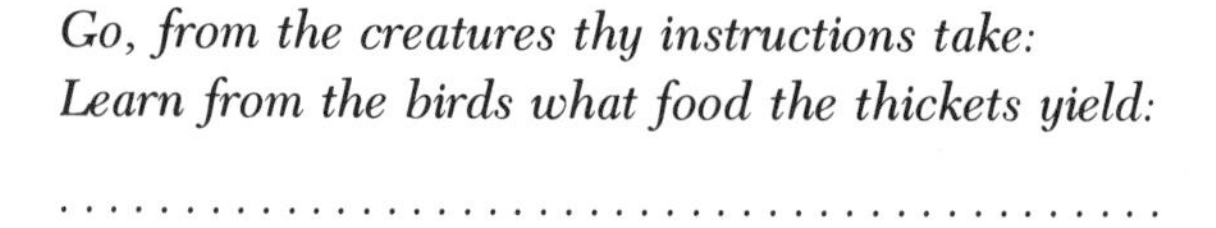

> *Go, from the creatures thy instructions take:*
> *Learn from the birds what food the thickets yield:*
>
> . . . .
>
> *Learn each small people's genius, policies,*
> *The ant's republic, and the realm of bees;*
> *How those in common all their wealth bestow;*
> *And anarchy without confusion know:*
> *And these forever tho' a monarch reign,*
> *Their sep'rate cells and properties maintain.*

It is interesting to note that as early as the mid-eighteenth century the poet saw the two essentials of the ants' social system as revealed by modern science, namely common property and the division of control between queen and workers. It is difficult to know whether Pope hit on this stanza by studying ants in his garden at Chiswick or by pure coincidence. It was probably the latter—a case, once again, of nature imitating art.

The collective nature of the ant system led William M. Wheeler about 1910 to put forward the idea that the colony itself was the animal and that the different kinds and sizes of ants found in the nest were like the differentiated cells in the body of a metazoan (many-celled, as distinct from a protozoan, or single-celled) animal. The colony as an animal also greatly appealed to the modern entomologist Rémy Chauvin. A solitary ant, he found, was incredibly stupid, but collectively the colony showed intelligence and skill. The ant brain by itself was minute, but if the brains in a nest of two hundred thousand individuals, for example, were taken together, they became an organ of reasonable size, capable of solving considerable problems. Chauvin thought the same about bees.

Regarding the individual ant as an animal and the nest as a social collection, or thinking of the nest itself as the animal, is not important. It is only a question of the meaning we attach to the word *animal.* What is important is that some minute creatures with very limited physical facilities have devised a system that enables them to survive for millions of years and to play a role in the ecology of the world. The system has overcome potential disasters such as fire, flood, dependence on slavery, and DDT, but it is by no means perfect. As we saw, it lacks complete economic efficiency. A great point to be borne in mind is the limited capacity of the ant brain: the ant does not think as we do but, like the bedbug, "gets there just the same."

It is difficult for us to imagine the manner in which certain animals know the world. We know it by means of our five senses—sight and the knowledge of color being the most useful, smell being of little importance. For dogs the reverse is the case. The electric eel lives in a world where changes in the electric field around it are detected, changes that mean nothing to us. For bees, and probably for ants, too, ultraviolet light is sensed, as is also the direction of polarized light; both are important to them but are of no use to us unless we have instruments to detect them. Wood ants spend at least half their lives in total darkness and there carry out a whole series of complicated functions, not the least important one being the exchange of food from ant to ant. Other things done in the dark are the laying and transporting of eggs, the raising of young, and cleaning the nest. The organs used to obtain the necessary information are the antennae, palps, forelegs, and certain sensitive hairs and bristles on different parts of the body. As H. G. Wells pointed out, in the country of the blind the one-eyed man is not necessarily king. Foragers can work on the darkest night, bringing home nearly the same food as they do during sunlight. Each nest has its own particular scent and it is because they have the wrong smell that strangers are detected. Nevertheless, ants are "capable of being deceived," as Wheeler wrote. He described the many creatures found in ants' nests as "this perplexing assemblage of assassins, scavengers, guests, commensals and parasites" and gave them a lot of elaborate Greek names.

Wood ants have a considerable memory. R. Jander concluded that in foraging they learn the routes in both outward and inward directions and also note landmarks, the memory of which persists for at least five days. Certain memories persist over winter, too, so that in spring the old sources of food can rapidly be exploited; they do not have to be rediscovered.

A feature of most ants' nests is the considerable degree of specialization occurring in them. Not only can the workers vary in size and structure or physiology, but workers of the same size dedicate themselves to special tasks for long periods of their lives. A worker's life lasts on an average about three years.

The newly hatched worker usually starts her career as a nurse ant, looking after the eggs, larvae, and pupae (the most precious articles the ants have). The following season she may perform a number of tasks, changing from one to the other as different needs arise. The main duties are foraging for food and for thatch (pine needles); building and repairing the nest, particularly the thatch; cleaning out the corpses and debris; sorting out food and other objects, such as grains of sand, bits of twig, and anything easily portable brought into the nest; removing eggs from the queen; nursing; feeding the larvae and helping adults emerge from the pupae; attending and feeding the queen; guarding the entrances, and, finally, military service—making war on other colonies or resisting attack from other wood-ant nests or from other species of ants.

W. W. Watt, in a witty poem, took the view that ants had no lesson for us:

HAD ENOUGH? VOTE ANT.

If the human race is so damn stupid that in 2,000 years it hasn't had brains enough to appreciate the secret of happiness—contained in one simple sentence that you'd think any grammar school kid could understand—then it's time we dumped it down the nearest drain and let the ants have a chance. The sentence is the golden rule. *Eugene O'Neill in an interview in the Herald Tribune.*

Are you sick of Homo sapiens?
Have you had enough of humans?
And the Heavens
Of the Bevins
And the Stalins and the Trumans
Let the ants
Have a chance.

Formica sanguinea steals the pupae of other ants and makes them slaves,
And when they die it buries them in special graves.
The aged *Stigmatomma pallipes* worker creeps off to die alone
To avoid being eaten alive when its social utility is gone.
Leptothorax lives off the food of *Myrmica,* its neighbor;
Some leaf-building ants employ child labor.
Crematogaster, a small ant, has formed an alliance with a large brown *Camponatus,*
And they go foraging together for ant nectar and lotus.
"Honey-pot" ants loll around like Romans
While workers help them fill their fat abdomens.
Atta raises fungus in gardens; *Lasius americanus* keeps cattle;
But the Legionary Ant spends all its time in battle.

Says Julian Huxley . . . : "Ants are among the very few organisms other than man which go to war."

Close the book of human folly
On the last ignoble pages–
Let the ants
Have a chance!
It's a slogan for the ages!
But as usual, the record
Is checkered.

The New Yorker, November 1946

2

MARCH

IN the Adirondack Mountains, in a clearing in the woods around Thirteenth Lake were a number of colonies of wood ants of the *Formica rufa* group. Two nests there are of particular interest to this story: numbers A and G. Nest A was eight years old, and Nest G had started eighty-two years ago. Wood-ants' nests had been on the site for millions of years, long before man had penetrated the area. In fact, the creatures had originated in America, before the continents separated, and from there wood ants had spread over the whole Northern Hemisphere.

Nest A was monogynous; that is, it had only one queen, whom we shall call Regina I. Nest G was polygynous; that is, it had fourteen queens—Regina II through XV. Nest A had around eighty thousand workers, and the Nest G complex had about half a million. The workers, as is always the case with ants, were all female, but not fully developed ones, as has been mentioned in Chapter 1.

Nest A was built over a rotten pine stump. It was a mound two feet high and three feet in diameter at ground level. On a dull day in early March it appeared to be just a peculiar pile of pine needles mixed with a few twigs and dead leaves, but this was because there was no spring activity yet. Within the nest was a large central chamber connected by galleries to many other underground rooms going down three feet below soil level. The thatch of pine needles protected the nest from winter rains and snow. It never froze within the nest and, as the thatch spread well beyond the circle of the underground chambers, the nest never became waterlogged. In addition, drainage channels in the nest conducted surplus water to the bottom and into the subsoil.

The nest had a number of openings to the outside world, but these had all been blocked in the fall to guard against intrusion during the winter. The sentinels stationed at these entrances during the summer had been withdrawn to the mass of hibernating ants in the depths of the nest, but occasional sorties by scouts were made to the openings to see if all was well and that the entrances were firmly closed.

Each of the lower chambers of the nest contained a ball of ants slowly moving. A peculiar feature of this sphere was the maintenance of an open corridor to the center of the ball. This ensured two things—that air reached the queen and that all the insects were at about the same temperature and humidity. As the outer ants in the ball cooled, they were replaced by the inner ants that had heated up a little by the natural metabolism of their bodies.

The nest contained no food reserves. In winter the ants lived by reducing their metabolism to a very low level, by drawing on the meager supply of food that had been stored within their bodies at the end of the previous summer, and by eating any weak or dead ants in the colony.

The principal character in this story, selected because of the unusual fate that eventually befell her, as will be told in Chapter 8, is a worker from Nest A, whom we shall call Labora. She was six months old at the start of this narrative, about one-third inch long, and reddish brown in color, except for the last part of her abdomen (the gaster), which was large, broad, and a dull black. She had six long legs.

Labora was one of the initiator ants, not exactly a leader but a creature that set a good example by starting some activity and thus inducing other ants to follow and do the same thing.

Sentenced by her psychological makeup to be active in pursuing the needs of the colony, and thus nearly always at work, she had to induce other, less conscientious, companions to join her. It was as if she were to say, "Heaving this caterpillar back to the nest is much more fun than lolling around in the sun." It was an action much like that of Tom Sawyer, who subtly induced his friends to undertake his punishment task of painting the fence. Of course, Labora could not say or convey any such message, but she could, by her activity (rushing around and tapping antennae), draw the idlers' attention to the caterpillar in question and thus awaken the food-collection reaction in them.

Both the endurance and the strength displayed by ants are phenomenal. Ants can, if need arises, work continuously for twenty-four hours or more. As for strength, an ant can lift an object several times heavier than itself and carry it away, a circumstance that leads to problems of balance because it then becomes very top-heavy. The creature's six legs and gripping feet help overcome balance difficulties, and, even if ant and burden do fall over and become separated, usually the insect quickly regains control and seizes its load again. However, at times, after an upset or for no obvious cause, an ant leaves its object on the ground. Usually it is picked up a little later by another passing ant and carried toward the nest once more; even so, it may yet again be dropped and once again recovered by another worker. Ants often work together in moving large objects. For instance, two wood ants may pull a dead caterpillar ten times their own size up a slope, which to a human observer might seem similar to two men trying to drag a dead cow uphill with their teeth.

The ants' small size accounts for the phenomenon. Following an argument dating back to Galileo, let us imagine that a man holding up a dumbbell is reduced in size by the linear factor of one-hundredth, bringing him down close to the scale of an ant. The volume and weight of the dumbbell, which vary as the cube of the linear reduction, are now one-millionth of their original magnitudes. On the other hand, the cross-sections of the men's bones and sinews (and thus their supportive strength) vary as the square of the linear reduction and are now one ten-thousandth of their original magnitudes. Thus, our tiny man appears to be one hundred times stronger.

Labora had emerged from her pupal case the previous August

and had immediately joined a number of other ants helping new workers emerge from their cocoons. They had to be dug up and moved to a propitious site in one of the upper chambers, where a faint glimmer of light came down a gallery. Here Labora watched one of her fellows slit open the end of a cocoon, enabling the new adult within to start its emergence. She was soon doing likewise. But not all the pupating ants were in cocoons; some were "naked," that is, just within a thin pupal skin, and were more easily released. To the ants the white cocoons, sometimes mistakenly called ant eggs, are the most precious products they have.

Just before the winter set in, Labora had been outside the nest for short foraging runs. Returning one evening, she had become interested in the activity of a sentinel ant closing an entrance and had helped her in this task. The memory of how it was done remained with her throughout the winter. The temperature of the outside world and of the nest had been falling, and foraging for food was no longer possible. The eggs could not be kept warm enough to hatch, nor could the pupae develop. The queen continued to lay a few eggs into the late fall, which the workers duly removed, but they did not put them into the hatching chambers. Instead, they ate them. Similarly, the larvae were neglected and eventually the nurse ants ate them. As all the ants were continually exchanging food with one another the eggs and larvae were a food reserve for the whole nest—nothing was wasted.

Labora thus had her share of this supply and went down into the queen's chamber in the depths of the nest. She passed the winter in the soft, routine movement of the ant mass, slowly drawing on the food reserves in her body. On March 5 she journeyed from the center of the mass of ants in this chamber to the upper parts of the nest, and, remembering her sentinel activity, she traveled up the galleries to Entrance 14 to inspect the closure. It was 1:00 P.M. Entrance 14 was on the southwestern side of the nest; the day was sunny. The blocking of the gallery was still secure, but the temperature had risen a little and some sunlight filtered in through the barrier of crossed twigs. She started to open the tunnel and was soon on the outside of the nest, in bright sunlight.

The first thing she began to do was to clean herself, an involved and complicated process. The ant has six legs, two antennae, and two maxillary palps (in the mouth); the manipulation of all these ten appendages is an involved matter in any action the ant takes, even in walking. It is amazing what good coordination the creatures have. The pair of antennae are the chief sensing organ, by means of which the ant gets its information when in the dark (where over half its life is spent). Although the eyes are used in the light, the antennae are still important, because they give and receive recognition signals, particularly by means of the touch-smell sense—the rubbing of noses, so to speak. The ant is equipped with a special hook—the strigil—on both forelegs that greatly assists the cleaning process. The strigil is curved and furnished with bristles, the better to effect cleaning, as the limbs and antennae are drawn through it.

After their long winter's confinement, the ants had a great urge to clean themselves. How involved this operation was could be seen by watching Labora's toilet. A cat has five appendages, and its cleaning activities may be seen to be complicated, but they are as nothing when compared with those of an ant, with not only double the number of attachments but also with a far more flexible body. The ant has very pliant joints between head and thorax and thorax and abdomen; it can roll itself almost into a circle if the need arises, even better than a cat.

Labora first passed her left-hand antenna through the strigil and under the foreleg, then treated the right-hand one in the same way. Twisting her head around, she passed both antennae, together, through first one hook and then the other, the cleaning comb helping the operation. With her antennae clean she felt much better, because she relied on them for so much of her information. She paused a moment and waved the antennae in the air to sense any activity, beneficial or dangerous, going on around her. Satisfied that all was well, she continued the complicated process. She stroked her head with her forelegs, then the abdomen and thorax with the hind legs. She next bent her abdomen forward between her hind legs, making herself almost into a ball by bending her head forward and down, and then washed the tip of her abdomen (the gaster) with her mouth. She wrapped the fore and middle legs on each side around each other, thereby scraping off the dust. The two hind legs cleaned each other as they twisted together. Next, an involved position was taken up, one found only in ants. Labora's body was balanced on a tripod made up of the top of her gaster and the middle and hind leg of her left-hand side. In that position she passed the tarsus of her left foreleg and all three right-hand legs through her mouth; she then repeated the operation by forming a similar tripod with the gaster and two right-hand legs. Naturally, much dirt collected in Labora's mouth—in the infrabuccal pockets—from where it was later discarded as pellets. The sun was slowly warming Labora; it was agreeable, and she next basked immobile in it for five minutes, her temperature slowly rising.

An important difference between the lives of man and insect is that the former, within narrow limits, maintains his temperature at a fixed level, whereas insects and all so-called cold-blooded creatures take on the temperature of the surrounding environment. Most cold-blooded animals have very limited powers of changing the temperature of their bodies, and if a certain temperature is needed before some activity can take place (for example, laying eggs), such action will not occur until the neighborhood rises (or sinks) to that temperature.

Bees and ants (but ants only to a limited extent) do have powers over their winter environments. Bees store honey, eat it during the winter, agitate their wing muscles, and thus generate heat. Their metabolism keeps the hive warm. But worker wood ants have no wings and use other methods to secure some control, usually by insulating the nest with a thick thatch.

Warmed by the sun to a temperature of 52°F, Labora reentered the nest, loosely closed the entrance, and went down to her chamber. On her arrival, her companions immediately noticed her higher temperature. One of them stroked her forehead with her forelegs and antennae, and Labora offered a tiny drop of the food still remaining in her stomach. It, too, was at 52°F and was welcomed by the ant. Labora now started to get excited. In the pitch dark she began moving around the outside of the ball of ants and into and out of the living corridor, drawing attention to her temperature, which, naturally, was dropping as she lost heat to the walls of the nest and the other ants. This was a critical action, because she had very little fuel left in her body. If she used up all her food reserve she would surely die. There was no food left in the nest, for the few larvae, pupae, and dead ants had long ago been consumed. But the colony can afford the sacrifice of a certain number of its members. The important thing for it is to know when the temperature is rising outside; hence, it sends forth such "thermic messengers" as Labora. That they die after doing this task does not much matter. There is a parallel in human behavior. For instance, in warfare it is not so important to the nation (the nest) if ten thousand men die, provided that Hill 60 is captured.

After a minute of this excited activity Labora set off upward again and was followed by six other ants to whom she had, in effect, passed on the message, "It's warm again outside." She soon reached the partially closed entrance, quickly opened it, and led four workers to the exterior. The two other ants had branched off on the way and were opening another entrance, number 12. The sun was still shining, but the sudden change from pitch black to bright sun did not seem to distress the ants at all—a surprising thing to a human, because the ants' eyes have no automatic iris, as we have, opening and closing according to the intensity of the light.

Labora and her four companions paused a moment around Entrance 14 and then started cleaning themselves and each other. Ants are very clean animals, and these ants were no exception. Their aim was to remove all particles of earth and dust adhering to their limbs and bodies and to leave them covered with a thin layer of saliva containing an antimold substance. This natural fungicide would prevent mildew from growing on the ants in the very humid conditions inside the nest. Cleanliness is not a virtue but a necessity for the survival of the species.

After their toilet the ants sought a warm spot and soaked up the sun. Labora, however, moved around the opening, inspecting the thatch and noting spots where it had been damaged by winter storms. At one place she dislodged a dead sowbug that had been trapped in the pine needles. It was a welcome discovery, because the creature could be used as food. With her sharp eight-pointed mandibles she cut off some legs and then a portion of the sowbug's stomach and carried them back to Entrance 14. The four ants who had been sunning themselves were nowhere to be seen, for they had reentered the nest, carrying warmth and news of warmth into it. Labora moved on to Entrance 12, where two ants were still sunning themselves. They were very excited by the sight and smell of the food and returned with Labora to the dead bug, which they all started to cut up. They fed themselves first of all. Each ant packed the food into a special pouch (the infrabuccal pocket) in her mouth. This pouch acted as a filter, allowing only liquids and fine particles to pass. Large bits of food would have blocked the narrow passages leading to the proventricular valve at the entrance to the crop and would have prevented regurgitation. The infrabuccal pocket was lined with a thin, but strong, skin of cuticle and could be compressed by special muscles that squeezed the liquid out of the food and sucked it (with its wad of small particles) into the gut. When her infrabuccal pocket was full, the ant spat it out, even though it still contained some nourishment. When ejected inside the nest, such "waste" pellets had a part to play in its economy (see page 67). Other portions of the great prize were carried into the nest, down the galleries from Entrance 12, and into the main chamber. More ants were now coming up from the depths and they met Labora and her companions coming down. The food they had was eagerly seized and eaten. Soon these few fed ants could be seen, felt, or smelled as having food in their stomachs and thus were capable of trophallaxis, or food exchange. One dead sowbug was not a vast supply of food, but it was a start, and the message was spreading around the community that food was available again.

By the time the newly stimulated workers got to entrances 12 and 14, clouds had covered the sun. The pine needles on the southwest side were still warm and encouraging, although the heat taken back into the nest by these first adventurers was not very much. By 4:00 P.M. the entrances were again closed.

The next day, March 6, was rainy. Consequently, there was no opportunity for a repetition of the warming process but, nevertheless, the handful of ants that had been outside the previous day were able to take advantage of the new conditions. Humidity within the nest was of great importance; the atmosphere inside was nearly always close to the saturation point—about 95 percent relative humidity or more. The wood ants themselves transpired greatly and were unable to retain water in their bodies. The nest was so well protected from rain by thatch that at the end of the winter the humidity there had fallen to about 90 percent, a dangerous level because it would lead to a further fall in temperature instead of to the rise they all needed.

When fluids evaporate, they draw heat from the surrounding bodies. Spill some alcohol on your finger. As it evaporates it absorbs heat and your finger feels cold. The same thing happens inside the nest; water evaporates from the ants' bodies, from the food, and from the nest walls. The lower the humidity is in the nest, the faster is the evaporation, and the colder the ants and the structure become. The ants, by keeping the humidity high, are able to keep the temperature relatively high, too, because the evaporation rate, and thus its cooling effect, is reduced.

Another function of high humidity is that it prevents the nest from catching fire. A mass of damp vegetation, such as a wet haystack, can heat up so much, because of bacterial fermentation, that at first it smolders and then bursts into flame. Wood-ants' nests can contain much damp vegetation—thatch, food, and usually a rotting tree stump. In the past, many nests that did not keep their humidity high perished by overheating and even fire. But, as in past ages, genes for maintaining high humidity seem to have existed, or to have arisen by mutation. Such genes were automatically selected out, as those ants—those "survival machines"—carrying them survived in the offspring in greater numbers than did those ants not having such genes. Obviously, ants that did not have the "maintain high humidity" genes left fewer descendants, because more of their nests perished by fire or were disadvantaged by excessive heat. There exists among wood ants what seems to be a built-in instinct toward the maintenance of high humidity in the colonies. If during the summer the humidity of a nest drops dangerously low and the temperature starts to rise, the ants open up ventilation channels to cool the nest, which is just what a farmer does with an overheating haystack. These activities of both ant and man have arisen by the automatic selection of advantaging genes. Not that there is a "ventilate wet haystacks" gene built into man, but such action is part of the "use your head" gene.

It is quite possible that some forest fires are started by a large wood-ants' nest bursting into flames. It does not take much to start a blaze. In summer, do not smoke near wood-ant nests.

On March 6 entrances 12 and 14 were cautiously opened at 10:00 A.M. The pine needles were moved from the upper side to allow rain to trickle into the galleries. Some of the ants also collected drops of water running off the needles and carried them in their mouths into the nest. Trophallaxis enabled them to share the water with numerous other ants, some of it eventually reaching the queen herself.

Water collection can be a dangerous activity for an insect, because of surface tension. A small insect attempting to take water from a drop on a smooth surface can be held by the strong surface tension of the drop and be unable to release itself. Its struggles to escape may exhaust and kill the creature, or an enemy may get it. The ants in this nest took the water from the pine needles very cautiously and were safe.

More entrances were opened and more and more water was taken into the nest. By 5:00 P.M. the nest humidity was once more 95 percent.

On March 7 it was sunny again, and more and more ants were coming up to bask in the sun. They were not only minute heat-accumulators; many of them were builders, as well. After early foraging, the repair of the nest structure was the next most important task. During the winter, dust and mud, splashed onto the nest by rain, were washed into the thatch and upper layers of the nest. The builders' first action was to start moving this fine material out of the structure in order to open up the network of surface galleries again. They worked from the apex of the nest downward, burrowing under the lower layer of pine needles and pushing them upward. They left the coarse materials, small stones, clods of earth, and small twigs, as these formed supports for the galleries. The result was an outer spongy zone covering an inner area of coarser stuff, while right in the middle was the original rotting pine stump. Beneath all this was a mass of rotting vegetable material, full of galleries and chambers, which was the cool area in summer and the hibernating zone in winter. The nest construction permitted a considerable range of temperatures to exist at any one time, according to the weather and the movement of the sun; this enabled the nurse ants continually to move queens, eggs, larvae, and pupae to those galleries and chambers most suited to their particular stage of development. The nest was built of holes rather than of walls.

Toward the end of March the building squads had cleared most of the outer-layer galleries of fine material and were bringing in small stones, hard clods of earth, bits of pine cones, and similar material to support the passages in the central zone of the nest. They were also beginning to pay particular attention to the thatch itself. The builder ants, leaving by the exits on the upper part of the core (the foragers used the lower exits), collected pine needles nearby and brought them back to the nest, usually two ants to a needle. The needles were carried to the top of the mound and released; hence, they tended to fall at random down the sides of the ant heap, thus preserving its symmetrical shape. The builder ants were continually changing the needles of the thatch, pulling old needles out and carrying them to the top of the heap if they were still serviceable, or carting them away to the rubbish dump if they were too rotten to be of use. The resinous pine needles resisted decay for much longer than the leaves of broad-leaved trees would have and were, therefore, very suitable as thatch for the nests. The constant changing of the needles preserved the symmetry of the heap and the optimum quotient figure (volume divided by surface area); it also enabled the ants to detect the attack of enemies boring into the thatch and to deal with them at once. The work involved was tremendous. Just imagine reroofing your own house every ten days!

Small insects—such as thrips, springtails, and mites—were occasionally found in the lower layers of the thatch as the needles were pulled out, and these served as food for the colony. The builders immediately ate them and, after digestion, offered that food to any other ants of their own nest that they met.

Toward the end of March, food was becoming plentiful. As soon as Labora entered the nest she was smelled as bearing food and was stopped by solicitations to share it, which she always did, dropping solid food and regurgitating a savory drop of liquid from her stomach, which was eagerly seized by her fellows and certain other creatures as well, as we shall shortly see.

3

APRIL

IN spite of the scarcity in Nest A, a small amount of food was being exchanged as the temperature rose, which quickly spread the message that food was the major need of the colony. By April small numbers of foragers were going out, mostly one- or two-year-old workers who remembered advantageous routes from the previous summer. In the main, the ants left by the exits on the southern side of the nest. Three yards due southwest of the site a worker ant found a holly bush, which she started to explore. In the thickest part of it she came across an overwintering red admiral butterfly (*Pyrameis atalanta*). It was torpid from the cold but still alive. The red admiral is probably the strongest butterfly known. Although it usually moves south for the winter, some specimens do hibernate in the northern states—in houses, under roofing tiles, and, as in this case, in thick bushes. They are safe from birds because they remain still and look like dead leaves.

The ant started to explore this strange leaf, which gave out a very unusual smell for a bit of dead foliage. By careful examination she soon realized that there was a big protein meal here, that it was fresh meat and too big for her to tackle alone. She cut off a wing, which would help keep it immobile, and then a bit of the abdomen, which she carried in her jaws. She set off back to the nest to recruit help in securing that entire badly needed meal. She had no language, as have the bees in their dance, but used another technique. As soon as she entered the nest she started dashing from one ant to another in an excited way. At first she showed the other ants the bit of food in her jaws, but in the agitation it was soon dropped. The smell of the food persisted, however. The first initiator ant she met was Labora, who received and quickly understood the "come with me" signal. Labora turned on her back, neatly folded her legs along her body, and was bodily picked up by the explorer ant, held by her mandibles, and carried back to the holly bush and the red admiral. At the foot of the bush Labora was released and followed her companion to the butterfly, now beginning to wake up and seek escape. The two ants now quickly cut off the remaining three wings and then the legs, and the butterfly fell to the ground. There they cut off its head, then small bits of the abdomen to carry home.

Two factors enabled them to find their way home. One was the angle the track made to the sun; the other, the memory of what the route was like, including the texture—rough, smooth, damp, or dry. The ants, as it were, have a built-in clock and make allowances for the sun's movements. Labora, having been carried upside down to the holly bush, relied more on the sun than on remembering the appearance and feel of the pathway that was taken. She had a considerable sense of time and knew that she had come out on a southwest track, though not in a straight line, for the path curved and deviated from that bearing. However, her memory was generally that of a southwesterly direction with loops and curves from it all the way along. She had to reverse all these curves in order to find her way home. The two ants set off together. Labora's companion, having made the journey three times before, was much more certain and used her eyes, thus enabling her to make short-cuts across some of the loops in the track; but Labora, with only one outward journey, had to rely on remembering the twists and turns. When a yard from the nest, she became confused and was lost, but by searching up and down she first smelled the nest and then found a well-worn track she remembered from the previous summer.

Her companion had already started kinopsis agitation—a rushing about the nest to excite the ants—and Labora joined in. Soon both of them were off again to the holly bush, this time followed by three ants each. There were now eight ants around the butterfly, which was cut up into manageable pieces, some of which were, however, too big to be moved by one ant alone. Labora and another ant had cut off about half the thorax, meaning to take it back to the nest. Labora had a better idea of the direction than her companion. Both ants were tugging at the morsel, the worker in a roughly northeasterly direction. But the other ant was pulling the prize the opposite way, convinced, erroneously, that hers was the correct route. The piece of butterfly jerked around and up and down as the two ants struggled. Wood ants are very alert to movement, and the tossing around of the bit of butterfly attracted the attention of its original discoverer. She knew well the direction of the nest and joined in the struggle on Labora's side. The food morsel, slowly but gradually, began to move northeasterly along the track as the ill-informed ant refused to give up. She hung onto the scrap with all her strength and was bodily dragged toward the nest. Naturally, the food moved slowly and jerkily toward home. After a journey of seven inches, the recalcitrant ant momentarily relaxed her jaws and let go of the food, intending to seize it again with a better grip, but the two other ants, with the brake now taken off, jerked the morsel an inch along the track. For the recalcitrant ant, the food had to all intents and purposes disappeared into thin air. She stood still a moment, cleaned her antennae, waved them in the air to see if they gave any clue, then hurried back to the butterfly body to cut off another portion. To the human observer, the whole operation seemed to involve a great waste of effort, not at all in accordance with our accepted view of the efficiency of the ant heap. Wasteful or not, the ants get the food home. And John Compton, in their defense, points out that moving such large objects is difficult for any animal. He asks if we have ever seen a couple of amateur furniture movers trying to get a large wardrobe down stairs, which bears considerable resemblance to the hesitations and mistakes of ants! The lack of clear-cut cooperation among ants is another example of the inefficiency of the system and contrasts strikingly with that of bees. A worker bee returning to the hive, or natural nest, with news of food passes the information on to her colleagues by means of a dance performed on the surface of the honeycomb. The dance is a figure eight; its inclination from the vertical and its speed indicate the direction and distance of the discovery, according to von Frisch.

Every time Labora and the other ants who had been at the butterfly got back to the nest, they indulged in the kinopsis activity, the excitement leading more and more ants to indulge in some particular activity. Soon there were scores of ants tracking backward and forward to the holly bush. They stopped their work only as the sun set, not because they could not work in the dark but because at that time of year it was too cold.

Soon there was considerable foraging going on. Small insects of all kinds, such as springtails, chalcids, and thrips, as well as young spiders, were taken, together with the sweetish secretions from tree buds, particularly those of the sugar maple.

Once food has been secured, other important activities can begin. Wood ants are particularly concerned to keep the surface of the nest clean and regular shaped. During the winter dead leaves had fallen on Nest A from deciduous trees and bushes, and the wind, rain, and snow had caused irregularities in its surface. Soon ants were busy cutting up and removing the dead leaves to the debris areas away from the base of the nest. The workers removed these leaves from the surface because they could shelter enemies, such as predatory insects and spiders. A clear view over the whole terrain is, after all, advantageous to any defending force.

For instance, the cleaners on Nest A found an unwelcome scarab beetle starting to dig into the nest under a dead leaf, and they immediately attacked the would-be intruder. The beetle at once feigned dead, spreading out its legs and remaining immobile. Although wood ants are particularly observant of movement and rely on it in their struggles, this feint was not sufficient to cause them to stop their attack; the creature had an alien smell and the outspread limbs offered holds to the defenders. They pulled at them, first one way, then another, so that the beetle had to adopt other tactics. It had a hard, shiny carapace and drew its antennae and limbs within it, presenting a surface on which the ants could get no grip.

Suddenly the beetle started into life again and commenced burrowing into the thatch. This action threw off some of the ants, but the others hung on and managed to turn the beetle onto its back and get it out of the thatch. The invader then rolled to the bottom of the anthill and moved away, with a deliberate movement that looked, to the human eye, as though it said, "I didn't really want to get into that nest!" Sour grapes!

Other groups of ants were foraging for pine needles and bringing them back to the nest to fill any pits and irregularities on the surface. There is a sound reason for cleaning the thatch, as for everything ants do: it keeps the nest warm and dry in winter and cool in summer.

The making and maintenance of a 2-inch or deeper cover of pine needles is a laborious task. If the surface is regular and approaches a hemisphere, the best use is made of a scarce resource; that is, the labor needed to find, transport, and place the thatch. Consider a smallish nest with a capacity of, say, 10 gallons, half above and half below ground. A 10-gallon sphere is equal to 2,310 cubic inches and has a diameter of 16.4 inches. The hemisphere aboveground will have a surface area of 422 square inches and be about 8 inches high. If the nest were a cube of the same capacity, its sides would be 13.2 inches long, with a total surface area of 1,047 square inches. If, as before, half were aboveground and half below, the surface area of the aerial portion would be 523 square inches and 25 percent more work would be required to cover the same volume of nest. Similar reasoning applies to irregularities on the surface. Thus a hollow, say, 2 inches in diameter and 1 inch deep exposes a surface area of 6.8 square inches, whereas the undisturbed surface (almost a flat circle 2 inches in diameter) exposes only 3.1 square inches, a considerable saving of effort and material.

The ants do not reason this out; they determine that a regular surface is a better economic proposition than an irregular one, but come to this knowledge instinctively, because the forces of evolution over the millennia have favored the races that did keep more or less regular, spherical nests at the expense of the kinds of wood ants that did not. The ants with irregular nest surfaces just disappeared, for saving from 25 to 50 percent of your building costs is a considerable advantage in the struggle for survival in a highly competitive world. In point of fact, the nests were not exactly spherical; their tops tended to be conical, so as to throw off rain more easily. For this very reason, a thatched roof on one of our own houses has to be steeper than a tile or slate one. Some species of ants use stones as roofs, making nests beneath suitable ones.

More and more food and more and more heat began to come into Nest A; heat was as important as food. The foraging ants picked up heat in the spring sunshine so that both heat and food were taken into the nest, together leading to the resumption of activity throughout the whole colony. Labora continued her foraging and was one of the first ants to take food down to the queen. The warmth of her front legs and antennae as she offered a soft bit of dead insect was agreeable to Regina I, who accepted it and moved among her attendants. Labora was accompanied by six other foragers, and soon the heat they brought in, minute though it was, stimulated the workers surrounding the queen into action. The workers and Labora began cleaning the dormant queen, took food from the foragers, and regurgitated it for her. She eagerly accepted it. After half an hour of such activity Labora began to urge the queen to return to the upper chamber, doing this by running toward an upward-leading gallery, then coming back and inducing a worker to follow her by means of antennal tapping until half a dozen workers got the message. They then got behind the queen and pushed her body and legs toward the upward-leading tunnel. All this was done in the pitch dark. The manipulation of six legs and two antennae per ant in confined quarters was, as can readily be imagined, a complicated business. Labora and her companions first had to induce the queen to move, then they blocked all the exits from the hibernation chamber except the one they wanted her to use. Once started, Regina I was as anxious as they to return to the breeding chamber. She also retained some memory of the path she should take to reach her summer headquarters. Food was not yet plentiful, but more and more foraging ants were being recruited all the time, and the neighborhood was being scoured for supplies. This inevitably brought the ants from Nest A into contact with those from Nest G, but at this time of year no aggression between the nests developed. Each nest had its own tracks and thus its own territory. The ants from Nest A remembered these trails and hurried up and down their old ones in their search for food.

The main track to the pine trees was fairly broad—about six inches wide—and a stream of ants was always moving along it in a characteristic manner. An ant would move forward for, say, about ten seconds, stop for a second, and then move on again. Their maximum speed was ten feet per minute, about one-ninth of a mile per hour. When an outward-bound ant met a homing companion they would both stop, rear up their heads, and touch antennae. The great majority of the encounters were between workers from the same nest, who recognized each other by means of the nest smell. After the exchange of the recognition signal (a password) each ant would continue on her way. Encounters with wood ants from other nests were mostly with those that had strayed from their own tracks. In such a case the tapping of antennae indicated to both parties that an error had been made. The home ant would rear up its head in a threatening manner as the intruder hurried away, seeking to rejoin its own territory.

By mid-April the main track of Nest A had, at midday near the nest, about two hundred ants per square foot on it. That is about one hundred ants per foot run, half going out and the other half coming in. It was not heavy traffic, for it was still early in the year. The stream of ants obviously became reduced as the primary track split into secondary and tertiary ones, but further reductions all along the routes were caused by some of the ants losing their way. They just drifted off from the track and, as a result, explored the intertrack territory until they again came to a route they recognized. If anything useful was found in their wanderings the lost ants spread the news of it to their companions. When the discovery was substantial, say, a few young caterpillars, a subtrack was developed to that point until that supply of good food was exhausted.

It was important for the wood ants to be able to find their way, both outward to established food sources and homeward with supplies. It saved wasting resources on useless traveling if tracks were established to worthwhile hunting grounds. At the same time, a certain amount of "leakage" of ants along the way, say, one per foot run, was beneficial, because the drifters explored the intertrack areas, where occasionally (as we saw) food might be found.

It is particularly important that the ants should be able to find the homeward path, because the loss of a loaded ant is of more consequence to the colony than the loss of an outward-moving empty forager. The loaded ant had not only its load but also knowledge of the place where that food had been found and thus where similar food might again be found.

To find the way was not a difficult task on the broad main track because it was so well marked and used, and it tended to carry the nest smell and thus establish it as "Property of Nest A. Keep off." But the case was different on a little-used tertiary route with, say, four ants per foot run, or an exploratory expedition with no established track.

Ants overcome this difficulty in a number of ways, one being to use the sun as a compass. They note the general direction of the sun on their way out and know that the opposite direction must be taken on their way home. This system would land an ant back somewhere in the immediate vicinity of the nest, if not at the nest itself, where a few probings would lead it to the main track or to one it recognized and knew.

Using the sun as a compass naturally depends on sight. Wood-ant workers each have five eyes. Three simple eyes form a downward-pointing triangle on the front of the head, and on each side of the head there is a pair of bulging compound eyes. All the eyes are fixed and cannot be moved, as ours can. Consequently, to change its field of vision an ant has to move its head or whole body, usually the latter. The fixity of the eyes plays an important part in direction finding.

The three ocelli do not so much give an image to the ant as they indicate light and dark. The compound eyes are elaborate, each consisting of about six hundred separate lenses or facets in a honeycomb pattern, which cause the light received to fall on a sensitive rod in the base of the eye and so on to the optic nerve and brain. The faceted eye, bulging almost like a hemisphere, is particularly well adapted to perceive movement, but it does not give much of a still image of the surroundings. When something moves, successive facets are affected and stimulate activity. What stands still is not recognized as anything but part of the general scene, which is why so many small insects sham dead on being disturbed. They may be left alone unless the ant's other senses, smell or touch, identify them. One of the reasons that wood ants continually stop when traveling is to see what is moving in their immediate neighborhood. It could be an edible creature or an enemy. The faceted eye sees movement more easily when the creature is still, for when it is moving the whole landscape is also moving.

Labora was particularly skillful in the art of direction finding. An account of one of her many journeys will make it clear how she used her facets to that end. On April 17, at noon, she set out on a foraging trip along a subsidiary track running northeast into the pine woods. The sun was thus behind her, due south, and consequently she noted that it was shining only into the backward-facing facets of the bulge of her right-hand eye and not at all into her left-hand eye. Moreover, a few facets were very brightly illuminated because the sun shone directly into them, so that by keeping those facets lit her direction, over a short period, remained constant. Every time she altered course, different facets became strongly lit and, as she was moving on six legs and thus somewhat jerkily, that was happening all the time. As already noted, the eye was well adapted to registering movement; as she herself was moving, it was equivalent to the sun's moving in the same jerky fashion. Keeping steadily northeastward, a group of neighboring facets got the full strength of the sun as she moved along. But the track itself was only generally northeastward, for it curved this way and that as it went along, in order to avoid objects such as stones, rocks, useless plants, fir cones, and so on.

Let us examine Labora's behavior in one of these deviations.

At a certain point the northeastward route came to a large stone and the track turned due north for a foot to avoid it. This was a considerable shift in the light pattern on her eyes. The backward-facing facets of both eyes were now lit. After going a foot in this direction the track turned due east for another foot and then reestablished the general northeasterly direction. The pattern on the compound eyes thus changed again. When Labora was going due east the left eye was unlit and the center facets of the right eye were fully illuminated. Labora and her fellow workers remembered this pattern and reversed it for the return trip. It was a pretty remarkable achievement for such a small brain to accomplish. Nevertheless, it is the sort of thing a fairly simple computer could do. If the eye were round, 1.5mm in diameter, and the facets were numbered from 1 to 600, there would be 28 of them across each diameter. The program would then be:

JOURNEY NO.	FULL ILLUMINATION ON FOUR FACETS, NOS.	DURATION, MINUTES	RESULT
1	Right eye. 59, 60, 80, 81	2	Track NE
2	Right eye. 55, 56, 76, 77 Left eye. 55, 56, 76, 77	0.5	Track North
3	Right eye. 45, 46, 65, 66	0.5	Track East
4	Right eye. 59, 60, 80, 81	indefinite	Track NE

The machine would reverse the signal numbers for the return journey, which is just what the ant does.

In our terms, Labora had set off northeast with the sun behind her at an angle of 135° to her general line of progress. When she moved north and then east to avoid the stone, the sun was at first directly behind her (at 180°), then at right angles (90°) to her line of progress. Naturally, she was quite unaware of these measurements, but she knew that on her homeward trip, if she turned around at once, she would have to reverse all those actions, and, in general, the sun would have to strike a certain number of forward-facing lenses of her left eye if she were to get home.

There are, however, two additional complications: the sun does not stay still, and it is at varying heights above the horizon at different times of the day and year. Labora and her companions were aware of the sun's movement and adjusted their return course to allow for it.

The sun moves 360° in twenty-four hours, that is, 15° per hour. Labora wandered from the track and spent an hour and twenty minutes exploring new territory but captured only a small springtail by the end of it. With her capture firmly in her jaws, she set off home, not on a line with the sun at 45° (the opposite of her outward track: $180° - 135° = 45°$) but with it at 25°, because from the state of tiredness in her muscles, and a sense of time in her brain, she knew that while she was foraging the sun had moved westward. In fact, it had moved 20°, and thus her new track should be $45° - 20° = 25°$ to her line of progress. Her return southwestward was thus achieved by her keeping a certain group of forward-facing facets illuminated on her left eye.

The acquisition of genes leading to this navigational skill no doubt took a long time. The ants have had a long time.

The next point to be made is more controversial. General considerations (including the navigation of the monarch butterfly on its long migrations) and some of my own observations on ants suggest that the height of the sun above the horizon gave a certain amount of information to Labora. It not only told her the time of day, confirming the information she already had from her built-in clock, but it also could indicate the time of the year; in fact, be a calendar, in our terms.

With a bulging, almost hemispherical, eye, only a small group of facets on the upward-facing surface got the full force of the light. The higher on the eye the group was, the nearer was the time to noon, always supposing she was on level ground, and if the maximum height (the zenith) was a little higher each successive day, then the nearer she was to midsummer. While it may seem strange that such a small creature as Labora could appreciate those differences, it must be remembered that many seemingly impossible things are quite usual in the natural world. Swallows fly three thousand miles to a winter home and in the spring find their way back to their own nest in the north. Even flowers can measure the length of daylight (or of night; it amounts to the same thing). For example, chrysanthemums flower only when the days are shortening in late summer and fall. That the messages conveyed to Labora's brain by the position of maximum illumination of her eye facets indicated both the time of day and of the year does, to my mind, not imply the possession of any miraculous powers, but only that Labora could remember and compare the movements on her eye facets. Ants remember many things—smells, routes, and food sources, for instance; they could easily remember sun positions as well. Of course they do not know the exact date (just the season), giving them a clue to future activities. Warmth and a rising sun mean, as it were, "Look out for aphids." Cold and a sinking sun—"Prepare for winter."

It was not very important to Labora to know when night was coming on, because she could forage then almost as well as during daylight, orienting herself by the moon or a bright star in a given constellation. Such knowledge was not valueless, however, for it made a difference to the hunting grounds. Usually, different game was collected at night from that secured during the day.

Using the elevation of the sun as a calendar was more difficult for the ant, for, though her eye was well adapted for the purpose, the height at any particular time of day varied with the date. For instance, if at Thirteenth Lake (43° N) the sun was at 47° over the horizon, it could have been at noon on either of the equinoxes (March 21 or September 21) or at varying times of day at any date between those two points in the calendar. For example, the sun would be at 47° over the horizon on June 21 (midsummer's day) at both 10:50 A.M. and 5:48 P.M. The ants would have readily distinguished the morning from the afternoon by the length of elapsed daylight, which may have served to orient them, because a measurement of the time elapsed from sunrise and the elevation reached pinpoints the time and date. Taking two of the dates mentioned, March 21 and June 21, in the first case the sun rises at 6:03 A.M. and reaches 47° elevation at noon (in just under six hours), then starts to sink again, whereas in the second case the sun rises at 4:20 A.M., reaches 47° elevation in the same time (six hours), but continues to rise until noon.

It is surprising what involved problems can be solved with simple observations, but at present we do not understand the physiology of biological clocks, and we really cannot explain the ants' use of them until we do. It might well be that the wood ant just has, not an eight-day, but a six-month clock and needs no further confirmation from the rise and fall of zenith suns. After all, much of a worker's life is spent in the dark.

Of course, if Labora, in her foraging stage, knew the season with some degree of accuracy, then she would also know that, with the sun at that particular height, it was either midmorning or midafternoon. Strangely enough, she *did* have a rough idea of the date, from the condition of the nest, the vegetation around her, and the kinds and states of the insects found in her territory. As to whether that particular elevation indicated morning or afternoon—no problem. The duration of daylight, of which she was well aware, informed her on that point.

All this was fine when the sun was shining, but wood ants cannot afford to be idle on dull days either. They have four methods of overcoming this difficulty. First, they can orient themselves on some prominent object, such as a building or a distant tree—for example, a wild cherry in blossom. Ants have to identify such a reference point, particularly if it is distant, because such a fixed object will not move relative to the nest, as will the sun. Ants do not have good vision for immobile things. It is thus possible that the distant flowering cherry could be mistaken for a hazy sun and allowances be made for its movement, when in fact it stayed still, a procedure that could lead to confusion and misdirection on return journeys. However, more light comes from even a hazy sun than from that reflected from the blossom, a fact helping ants distinguish between the two objects. It is not likely that errors of this sort occur with any frequency.

The second sunless day method works on a dull day with patches of blue sky or very thin cloud, when ants could orient themselves on the polarized light reflected from such patches. Such light is traveling all in one plane, and ants could sense this plane of light and use it as a reference device. Moreover, they could allow for its turning as the sun moved.

The third method wood ants have of orienting themselves on sunless days uses their ability to register ultraviolet light, which we cannot do because our cornea filters it out, but those of us who have undergone a cataract operation are said to be sensitive to it. Some ultraviolet rays pass through the cloud layer so that ants could "see" the sun even under those conditions, unless the clouds were very black and heavy.

The fourth method uses ants' good visual memory for features of the track, helping them remember details such as the shape of pebbles, cracks, and blades of grass. Labora had a good visual memory for such details and was able to use several methods at about the same time, one succeeding the other as conditions varied. She frequently took shortcuts on her way home. This meant that she let her sense of direction (sun or polarized-light orientation) take precedence over her visual memory.

Within Nest G, activities were very similar to those in Nest A. The fourteen queens were being returned to the egg-laying chambers. Eggs were being produced and moved to the brood cavities under the care of nurse ants, who cleaned the eggs and took care of the larvae, constantly moving them about the nest as conditions altered. The nurses tended to be the ants born in the late summer or early fall of the previous year; most of them had never been outside the nest. The queens produced a secretion liked by their workers, thus ensuring their receiving careful attention. Instinct and appetite coincided. It was not a case of loyalty to the ant nation or of intelligent anticipation of favors to come in preserving the ant species but was once again the working of natural selection. Those strains of ants that carefully tended the queens, and those queens that strengthened this care by providing an agreeable condiment, survived at the expense of the races that did not excrete the characteristic ant "soma."[4]

The larvae, we saw, were constantly moved about the nest to be licked and cleaned. They, in turn, produced a chemical complex that influenced the queens. It was a male-inhibiting pheromone, ensuring that only genetically fertile eggs were laid; that is, eggs that would produce only larvae growing to female ants in due course. As the eggs passed down the queen's oviduct, a few sperm were squeezed onto them from the store in the queen's seminal vesicle. The sperm entered the egg by the micropyle (a minute opening in the eggshell) and one spermatozoon fertilized each egg. Fertile eggs produced females and the routine standard feeding ensured that the females born were imperfect—that is, that they would be workers, the prime need of the nest for survival at that time, and not potential queens. By the end of April a steady flow of fertile eggs was being processed by the old workers and the newborn recruits.

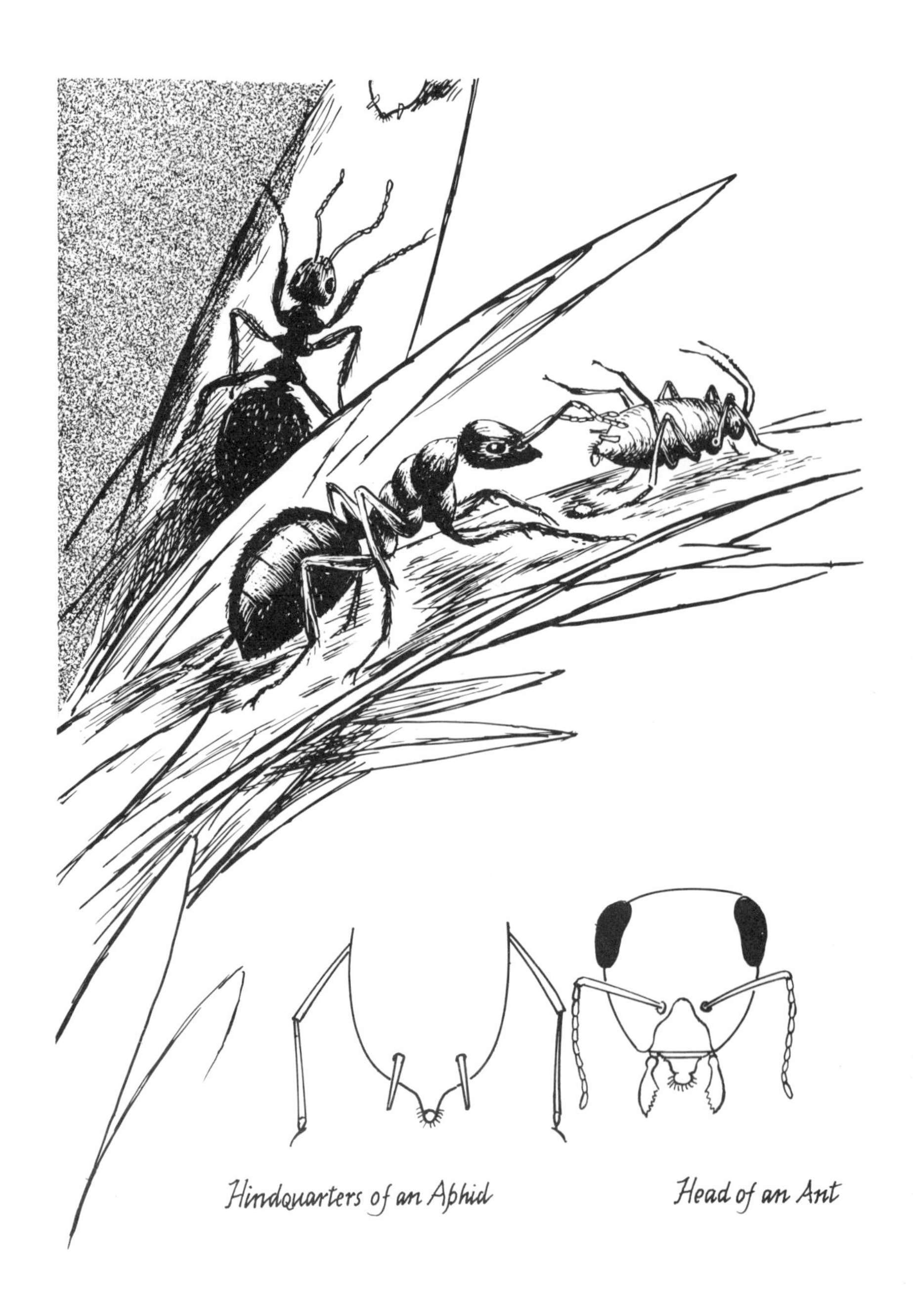

Hindquarters of an Aphid

Head of an Ant

4

MAY

NEST A was on the borders of a clearing in some natural forest, mostly pines and eastern spruce, but with a few hardwoods and maples as well. Aphids, such as the spruce gall aphid (*Adelges abietis*), were now hatching in considerable numbers among the spruce needles and on the pines. Labora was particularly concerned with this activity on a tree eleven yards along the main track, on the south side: Tree 6. On April 28 she had located a few aphid eggs and had paid them several visits, licking and cleaning them on subsequent days. On May 1 three eggs hatched, and the ant carefully carried two of the young nymphs, each to a young spruce needle. The third youngster found its own leaf. Each aphid immediately sank its delicate beak into the leaf and started to suck the sap. The aphids grew rapidly and after two days molted, casting their first skin. Labora guarded them during this activity, mostly by constantly moving up and down among the needles and frightening away spiders and ladybirds, like a shepherd keeping off wolves. After another four days, the aphids were fully grown and giving birth to living young. Each insect was taking up considerable quantities of sap, and it was this liquid that was of interest to Labora.

Aphids can be damaging to trees, not so much because of the sap they withdraw and use for themselves and ants as because of the virus diseases they can carry and introduce into their hosts, or the deformations of growth that the saliva they inject can cause in the vegetation. Some aphids push their mouths into the phloem of their host plants and suck the processed sap circulating in the sieve tubes. Such insects inject no saliva, or almost none, and do not affect the plant (M. J. Way). Other aphids feed on the parenchyma, the cells themselves, inject an almost continuous stream of toxic saliva into the plant, and can cause much damage.

Most of the aphids fostered by the wood ants were of the former type and did little harm to the trees, although the spruce gall aphid, as its name suggests, did cause a little deformation of the needles. However, such small damage as they did was more than counterbalanced by the tree-feeding caterpillars and beetles the ants killed. Forests with about two wood-ant nests per acre are always healthier than those with none.

Aphids are greedy feeders and suck up much more sap than they can use. The surplus sap, almost unchanged, is excreted from time to time. When no ants are present, an aphid forcibly ejects a drop of sap from her anus so that it falls a little distance from her, usually spreading over a leaf beneath her position. The sap is a good culture medium for certain fungi, and leaves coated with it soon have a sooty mold growing on them, turning their surfaces quite black, lowering their photosynthetic power and hence the general well-being of the plant. The ejected sap is a rich ant food and, according to the poet,[5] also fed Kubla Khan. The secretion contains many free amino acids, amides, proteins, minerals, B vitamins, and sugars—a source of energy for the ever-active ants. Consequently, wood ants take great care of aphids and collect the honeydew.

Aphids usually have two cornicles pointing backward on the back of the abdomen, from which they can eject a sticky and repellent fluid. The ejection of a drop of this fluid can stick together the jaws of an attacking ladybird and lead to its abandoning the hunt.

But most of the species of aphids (there were at least sixty-five of them) attended by the wood ants did not have cornicles, or had very reduced ones. The aphids had come to rely on the ants as protectors against their enemies and, in effect, paid for this service by supplying the ants with honeydew. The aphids, apparently such soft and vulnerable insects, employ other defensive devices as well. One of their tricks is for all the greenflies to kick their hind legs in unison, frightening away an aggressor. Some aphids, usually big ones, even drop off a tree altogether when threatened. To the attacker the food seems mysteriously to have vanished. The danger past, the greenflies gently climb back up the tree again. The spruce aphids kick, but not at their friends the ants, nor do they indulge in the vanishing act.

Inexperienced ants, such as Labora in her early foraging sallies, often make mistakes in their first contacts with aphids, as can be seen by following some of Labora's feeding activities. In or on the nest she solicited food from a companion first by facing her, but sight played no essential part in the transaction, because it could take place in the pitch darkness of the nest as well as outside. In the dark the first thing to be done was for Labora to establish that she was in contact with another ant. This was soon discovered by palping the object before her with her forelegs and antennae. If it were an ant, this was confirmed when the antennae touched, at which moment the donor ant also knew she was in touch with another of her kind. The palping, in addition, told Labora that the donor ant had food to offer. The smell also informed both of them that the respective creatures were ants from their own nest. Labora then had to get into the right position to receive any food offered, doing this by using her antennae and the labial palps of her mouth. The donor ant also used her palps to help get the position right. Once the donor ant had established the correct position, Labora started to stroke her head with her forelegs; the donor returned the compliment, but with fewer and shorter strokings. She then regurgitated a drop of fluid from her stomach, which Labora slowly absorbed, breaking off the contact when the fluid had all been taken in. The transaction was carried out more quickly in daylight, first, because sight was used for recognition, and second, because it was quickly followed up by using the senses of touch and smell.

On May 6 Labora was on guard duty among the multiplying spruce aphids. Suddenly she saw what appeared to be the head of an ant immediately in front of her and at the same time smelled food. Having been active for some time, she was now hungry, so she approached the apparent head of a companion ant, intending to obtain a drop of food from her. The curious thing is that the rear of an aphid, with small cornicles, looks somewhat like the head of an ant (M. J. Way). Moreover, the aphid assists this illusion by raising its hind legs in the air so that they thus appear to be the antennae of an ant. The aphid's cornicles gave the illusion of being the ant's mandibles (jaws), and the anus the mouth parts. As soon as Labora palped the object in front of her and touched the aphid's hind legs with her antennae, she realized that she had made a mistake, for she got no answering response: there was no ant in front of her. But at this moment the aphid started to excrete honeydew, which was held on a few hairs arranged in a circle around the anus for the purpose of retaining the drop while the ant absorbed it. Labora started at once to suck up this food so conveniently offered to her. The reaction to the availability of honeydew overrode the trophallactic (food-exchanging) impulse with which she had started. After a few mistakes of this nature, she soon learned how to distinguish between the rear of an aphid and the head of an ant and became an active collector of honeydew.

To confuse those two objects may seem almost impossible to us, but we ourselves do not always see differences between roughly similar things until we get well acquainted with them. For instance, a city dweller, born and bred, once asked a countryman, a tough old fruit farmer, "Mr. Shorter, how in winter do you tell the difference between a plum and an apple tree?" "Tell a horse from a cow, can't you?" was the scornful, laconic reply.

Labora continually made journeys to and from the tree, always loaded on her return with that important food taken from the greenflies. She always went back to the same few leaves containing "her" greenflies—her territory. During daylight she used her eyes to find the way and at night her kinesthetic sense. (This sense provides a remembrance of turns and distances traveled. Submarines, submerged for weeks at a time, navigate by using much the same system.) But at night, unless it was very warm, Labora did not milk the aphids much. She had but to stroke the abdomen of one and it immediately started to excrete a drop of the precious fluid, unless, of course, that particular aphid had produced one a few seconds previously for another ant. The aphids put out from five to fifty droplets each ten hours, depending on temperature, the time of year, and the sap pressure in the tree.

The similarity of this action to the keeping and milking of cows by humans can hardly escape the notice of the reader and has been commented upon by many authors. However, ants do not herd their cattle; it is as if their kine were permanently tethered, because once the insect has inserted its beak into a good feeding point it rarely moves. But it will change its position a short distance if conditions become too crowded.

Labora became quite skilled in locating aphids with a supply of juice available and thus saved time by not "milking" insects that had just given their all to another. The fullness of a greenfly was shown by its swollen abdomen and a certain amount of honeydew oozing from the anus.

In May, honeydew was an important item in the feeding of Nest A, accounting for over 60 percent of the total food taken in. Tree 6 alone supplied considerable quantities. A fully grown adult aphid put out ten milligrams a day, and a first instar about half that amount—a surprising quantity, because that figure was about one and a third times its own weight. The younger and smaller aphids produced smaller drops but more of them. The adults, besides producing young, excreted about a third of their weight in honeydew every ten hours (the active day). Labora and her companions carried all this back to the nest. Fourteen spruce and pines were being used by the twenty thousand foragers in Nest A, who were thus taking home about ten pounds of honeydew a week. The Nest A ants obviously regarded the fourteen trees near their nest as their property and repelled all intruders, particularly ladybird adults and larvae, syrphid flies, and ichneumon parasites. They also had to keep off scouts from the nearby Nest G.

On May 12 a G worker on such an expedition had succeeded in soliciting food at the base of Tree 6 from a Nest A worker low in cautionary genes. In this A ant the desire to supply the nest with an attractive piece of food—for at this time of year the demand from the nest was almost unlimited—overrode her weak "recognize and oppose a stranger" characteristic. We could see it as greed overcoming caution. Realizing that she had just been given a drop of the much needed honeydew, the intruder started to move up the trunk to find the source. She met Labora coming down, walking in the customary jerky way. The two ants raised their heads and forelegs in the usual manner on meeting and touched antennae; the effect was startling. Labora knew at once that an intruder ant was trying to enter one of Nest A's properties, and she immediately attacked. Ordinarily, the G ant would have retreated when threatened, but the prospect of securing the much prized honeydew overrode the retreat reaction and battle was joined. Labora planted her left foreleg inside her opponent's right foreleg and, at the same time, held down G's left front leg with her fore and middle legs. She then tried to lift her left foreleg so as to turn her opponent over. Labora had the advantage of position and weight; she was above the G ant and much heavier because her abdomen was swollen with honeydew. But the G ant could not be defeated by such a simple maneuver. She rode up over Labora's head and thorax and attempted to bite through the petiole and sever the abdomen. Labora was now in difficulties, gave the danger signal by excreting the "alarm—danger" substance, and, with a twist and a great heave, pushed her opponent's dangerous jaws away. Both ants fell to the ground, by chance onto a bare patch of land next to the main track, which was fortunate for Labora, because she was giving off the alarm smell and thus attracted the attention of an outward-moving worker from Nest A, who happened to be passing. On the flat ground the G ant was beginning to get the better of Labora, whose movements were restricted by her full stomach, but the situation changed when the worker from Nest A arrived. The first thing she had to do was to discover which was friend and which foe; a few antennal touches sufficed for that. The rescuer now seized a hind leg of the intruder, pulling it out and twisting it at the same time, thus exposing the thin skin at the femur joint of the enemy ant—a sensitive area. The rescuing ant now reared up, bent her abdomen around and forward, and from her anus ejected a thin stream of formic acid at the joint. In three seconds' time the leg was paralyzed. Another Nest A ant now joined in and she and Labora immobilized two more of the intruder's legs. The intruder was now doomed and soon died. Labora shook and cleaned herself, hunted around until she found the track, and started back to the nest. A small amount of the alarm substance still hung about her, so that she received considerable attention from the outward-bound ants, but after a few inches the chemical had all evaporated and the rest of her journey was uneventful. As usual, she proceeded in a jerky manner, pausing every now and then to survey the surrounding scene from a stationary point and to regurgitate a drop of honeydew for the encouragement of the outward-going ants.

At the scene of the struggle the two rescuing ants cut up the visitor from Nest G and carried the succulent parts back to their colony.

Labora worked hard on the honeydew exercise, being particularly vigilant in keeping ladybirds, lacewing flies (syrphids), and aphid wasps (Pemphredoninae) away from the greenflies. She frequently spent all night on Tree 6, when spiders were the main threat. The result of these attentions was that the aphids multiplied excessively, leading to the production of winged forms and the forced migration of the ever-increasing numbers of young. Uncollected honeydew began to fall from the pine needles, attracting bees and flies. The Nest A foragers took no notice of the bees. In the first place, they were too big to tackle, and second, they did not take the excretion directly from the greenflies but only sucked up the excess honeydew spilled onto the leaves. On the whole, the bees performed a useful service, for by taking up the surplus syrup they kept the leaves clean and thus free of the inhibiting sooty mold fungus, enabling the ants to continue drawing their nectar from clean, healthy trees. The flies were chased away by the constant activity of the foragers, as were the green lacewings (Chrysopidae, of the order Neuroptera), avid feeders on greenflies.

The foragers now started to kill aphids, particularly the young migrants and the winged forms, and take them back to the nest. The flesh had a rather higher protein content than the honeydew. Greenflies are juicy creatures, and their consumption in the nest also helped maintain the high humidity needed there. In mid-May the Nest A queen was becoming very active and laying a great number of eggs every day, making big demands on protein reserves, so that young aphids were a welcome addition to the diet. The parallel with the human herdsman's behavior in culling animals (deer, for instance) to prevent their becoming too numerous for his good—and frequently for theirs—is obvious. The ant operation, however, is not a thought-out process but an automatic reaction to the movement of the insects, which in itself is induced by crowding.

On the evening of May 26 Labora returned to the nest and spent the night there. By chance, on May 27 she left the nest by one of the upper exits on the eastern slope, usually used only by the builder ants. It was a bright, sunny morning and as she moved down the slope she came across three workers engaged in cleaning each other and sunning themselves. Initiator as she was, the activity engaged her attention and she joined the group. They exchanged food with one another, pulled the antennae through their strigils, exercised their limbs one after the other, and lay around in the morning sunshine, showing every appearance of enjoying themselves. A young worker—a nurse ant—approached Labora and solicited food, which she provided. The worker then tapped Labora's antennae, opened her jaws, and pushed them against Labora's right foreleg, but did not close her jaws or seize the leg. It was an invitation to a mock fight. Labora pushed her head under the young ant's thorax, lifted her up, and turned her over, but she did not follow this up by grabbing and stretching a leg. The second ant then did much the same thing with Labora, and so the exercise continued for half a minute, the two ants tumbling around and rolling down the heap, seriously disturbing the ascent of some thatchers carrying pine needles. At the bottom of the heap Labora entered the nest. Her physiological condition was now such that she was ready for a change of occupation. Having entered the nest, she proceeded to one of the lower chambers where the sorter ants were at work, almost in the dark except for a faint ray of light coming down a gallery and a slight glow from the phosphorescence of decaying vegetation in the walls. She was soon working with the others.

The sorter ants were particularly sensitive to the requirements of the colony and were always breaking off their activities to exchange food, particularly with the queen's attendants, for it was only in this way that they could sense what ingredients were needed. The foraging ants came in, day and night, with a steady stream of objects, honeydew being the most usual at this time of year. The sorters knew the nest could use all the honeydew it could get, so that sated foragers were pushed upward toward the brood and the queen's areas, where they gave the food to the royal attendants, who passed some of it on to the queen. Trophallaxis also took place between the foragers and the sorters for the latter to check that the entrants were in fact carrying honeydew and that it was of the required quality.[6]

But in addition to honeydew, the foragers brought in considerable quantities of other materials. In fact, they tended to pick up any small portable object and bring it along. The senses of touch and taste in the antennae and the taste of food passed in the food-exchange drops told the sorters what was available and what was needed. The sorters' role is thus vital to the colony. Small stones, coarse sand, and lumps of earth are passed to the builders for use in constructing and strengthening galleries.

If the supply was excessive, the builders simply carried the surplus out again to the dumps at the base of the mound. Bits of old fir cones were quite common and were also given to the builders and thatchers, though the seeds, when present, would first be pulled out for food. A number of early seeds were also brought into the nest and were more carefully examined by the sorters. The day Labora joined them (May 27) seeds of goat's rue (*Tephrosia virginiana*) and wild indigo (*Baptisia tinctoria*) were arriving. Both were legumes, high in protein and useful for supplementing the queen's diet, but the sorters found the rue more to their taste (which was influenced by the makeup of the exchange food they were getting from their companions, as noted above) than the indigo, and they turned most of the latter seeds over to the scavengers. The rejects were carried to the base of the mound and dumped, where later in the year some of them germinated. Such growths in the past gave rise to the idea that ants practice true agriculture, but this is not the case. The "fields" are merely rejected seeds that have been dumped for some reason. The advantages of a wanted plant growing in or near the nest, objectively, are obvious so that in the course of time—a million years is nothing to ants—open-air agriculture might have evolved. Some ants—the leaf cutters—already practice fungus culture. They live on a fungus growing in nest chambers on chewed-up leaves brought into the nest by the foragers, doing much harm to tropical agriculture in the process.

The apparent waste of bringing in food and then carrying it out again is a myrmecan solution to a problem common to both ants and man—what is the best way to get supplies? The solution adopted by ants is what it is mainly because of difficulty in communication. The sorters and the nest in general have no way of saying to the foragers, "Do not bring in wild indigo," or of ordering, "More goat's rue, more aphids." To make sure of getting essential or desirable materials, a little of everything available has to be brought in, even if some is in excess of the needs of the moment. Although the foragers are subject to the taste effect of the communal food from the "social" stomach, two things override it. First, the reaction to small moving things (insects, for example) arouses the hunting reaction, and, similarly, portable objects are felt to be waiting to be picked up and carried away. Second, the diluting effect of fresh food eaten by the foragers, particularly honeydew, masks the social stomach instructions. For instance, if the latter was giving the signal, "More protein, less sugar," then an ant full of honeydew would not get the message—not that Nest A ever had occasion to dump any nectar.

The system produces results, and its inefficiency is perhaps fortunate for us, for had ants been able to talk or to communicate with each other even as well as the bees (see pages 39–40), the higher animals, including man, might never have arisen.

Another cadre in the nest is the cleaners, with whom the sorters work closely. The cleaners are responsible for hygiene; they respond quickly to the smell of dead adults, larvae, or pupae and remove them to rubbish dumps. Any undesirable object in the nest too big to be removed would be walled up with particles of earth, stones, and twigs. In Nest A the sorters passed their rejects to the cleaners who, conditioned to the removal function, at once carried the unwanted objects out to the garbage. Inefficiencies could occur here because a cleaner, in the way ants have, might put an object she was carrying, say, an unwanted seed, down for a moment. It could then be picked up by a forager and returned to the nest, quite likely to be once more rejected by the sorters.

On the spruce trees the foraging ants were finding a number of young caterpillars of the spruce budworm (*Choristoneura fumiferana*), and many were captured, not without a struggle, and carried back to the nest.

The budworm belongs to the Tortrix family of moths, a characteristic of which is the intense wriggling the caterpillar indulges in if touched. When a young budworm is seized by an ant it lets go with all its legs (six true legs and five pairs of abdominal, or prolegs), throws itself violently from side to side, and tries to drop from the needles on which it has been feeding. When it succeeds in this it spins a silk line from its mouth as it drops, first entangling it with the leaf on which it has been feeding. If the caterpillar shakes off its attacker it can ascend its lifeline and regain its feeding site. An attacked budworm frequently wriggles off the tuft of needles, but the ant nearly always manages to cling to it so that both fall to the ground, where the struggle continues, usually on less advantageous terms for the budworm. Once on the ground, the ant usually gets help because the violent movement attracts the attention of other foragers, who rush up, cut the silk lifeline to prevent escape that way, and seize the luckless caterpillar at several points, one particularly vulnerable one being just behind the head. Formic acid is sprayed into the wounds, and soon the budworm is dead or paralyzed. The prey is then dragged back to the nest, with the ants walking backward. Such activities greatly benefit the trees, for the budworm can be a serious pest and cause extensive defoliation.

The sorters, who were in constant touch with the queen's attendants, carrying the chemical messengers, did not reject the young budworms. By May 28 the queen was laying eggs and secreting pheromones at an ever-increasing rate. The existing staff was unable to cope with it, and recruitment was necessary from the ranks of the sorters and foragers. On May 29 a royal attendant approached Labora and offered her a drop of regurgitated food, which was accepted. The morsel contained a considerable quantity of the queen substances, which influenced Labora's behavior. When the newcomer tapped antennae and stroked Labora's face in a certain way, she rolled over, folded her legs, and, gripping the attendant with her jaws, was picked up. Now firmly held, she was carried upside down, beneath the attendant, to the queen's chamber and released. Attracted to the queen by her abundant body secretions—a sort of sweat—and the ample honeydew about her jaws, Labora joined the other ants licking and cleaning the queen's body. She had now, so to speak, joined the royal household.

Queen Regina I was then laying 200 eggs a day, working throughout the twenty-four hours. Though it was not her maximum output, which would be about 330 eggs a day, she started and ended the reproductive season (April to September) at much lower figures and stopped altogether in October. Nevertheless, it meant she was producing eggs at the rate of about 20,000 a year. At her peak, Regina I would be laying an egg about every four and a half minutes, which gave the workers around her plenty of time to remove the egg and clean and feed their monarch between each oviposition. Obviously, she needed a sizable staff to supply her food, transport the eggs to hatching chambers, remove feces, and keep her clean. The resulting larvae also made constant demands for nursing staff. All the larvae were removed several times a day to different sites, according to their age and the prevailing weather;

this was the kind of work done by Labora when she was a young adult.

Regina I had held in storage an immense supply of sperm within her, which she had obtained on her mating five years previously, but by now it was considerably reduced. It was kept in a special sac—the *receptaculum seminis*, or spermatotheca—and had to last the whole of her life, which could be ten to fifteen years. To keep the sperm alive all this time, Regina I had a pair of glands at the apex of her seminal sac secreting a nutritive fluid for that purpose. The queen had the option of laying fertilized or unfertilized eggs. In the first case, a few sperm were squeezed out of the *receptaculum seminis* as the egg passed from the ovaries down into the vagina, and one fertilized it, whereas in the second, no such action took place. The unfertilized eggs produced males, and the fertilized eggs produced workers or princesses (virgin females), according to the feeding given them during the larval stages. Far more eggs producing females were laid than eggs producing males, as workers were the prime need of the colony.

On May 30 Labora happened to be near the tip of the queen's gaster when an egg was laid. She gently took it in her jaws, put it down at a little distance, and started to clean it with her tongue; then she carried it to a warm gallery (about 72°F) and tucked it into a small pocket in its wall with a number of other eggs. It was minute, white, long, and oval and had to be very carefully handled.

5

JUNE

IN the early part of June, Labora devoted herself to the care of the queen, particularly in tending her eggs. When an egg was laid Labora first cleaned off dirt and any adhering body tissues, then carried it to one of the egg piles in the galleries set aside as hatcheries. She continually visited the piles, removing dirt or any mold that had started to grow on them. The constant licking left a coat of saliva that had a fungicidal action and prevented the growth of mildew. She also checked that the embryos within the egg were still alive and ate any eggs in which they were not. On June 11 the eggs in her pile began to hatch, but she was not there at the time. In fact, from that moment she lost interest in that particular pile. Responsibility for those larvae was now handed over to the nurse ants.

The young larvae stayed in the egg pile and were licked and cleaned by the nurses. Up to their first molt the larvae fed on other eggs in the batch. After molting, the young creatures—little fat grubs with no eyes or legs—were removed by the nurses and taken to a brood chamber, where they were fed with prepared meat. Caterpillars, aphids, and other animal foods were first skinned, then soaked in saliva and given to the larvae. The rejected pellets from the infrabuccal pockets were sometimes fed to them as well. Honeydew regurgitated from the nurses' crops was also supplied.

The special food given to the male grubs and the few larvae destined to become fully developed females was secreted from the labial glands in the thoraxes of the workers. It was particularly nourishing, being reminiscent of the "royal jelly" fed to bee larvae selected to become queens. The labial secretion was obviously much smaller in quantity than the amount of food brought up from an ant stomach. Consequently, the care of the royal brood demanded a considerable body of wet nurses, some three thousand workers in the case of Nest A. The queen herself also took some of the labial gland secretion, about two hundred workers being needed to supply her with that.

The workers in general and nurses in particular were devoted to the larvae and pupae, their first concern in times of danger being to secure their safety. Nevertheless, they did not seem to differentiate between eggs and first instar larvae, for the latter were not fed but licked and moved around in the egg pile in the same way as eggs. After the first molt the larvae were the subject of constant attention, and the young in their turn responded to the touch of their nurses. After contact had been made between a nurse and a larva, the latter would bend its head a little forward and down to receive food, but the main movement was from the nurse ant who, in the pitch dark, located the larva's mouth with her antennae and forelegs, and then offered it the food, either a protein morsel or a regurgitated drop of honeydew. A larva would not always accept the proffered meal. In the case of a refusal, the nurse moved on and offered it to another grub and eventually got rid of it. The larvae sometimes excreted a sugary substance that the worker ants liked and licked up. It was an action strengthening the already existing instinctive attachment to the service of the young. The larval excretion also contained chemicals influencing the behavior of the queen when the trophallaxis process eventually led to those substances reaching her, which could be quite soon. For instance, on one occasion when Labora delivered an egg to her particular egg pile she exchanged food with a nurse ant who had just been feeding a larva and had taken up some of the larval secretion from its mouth and skin. Labora then returned to the royal chamber and exchanged food with another royal attendant, who shortly afterward fed the queen. In just three removes the queen got the chemical message emanating from a larva. Power was thus nicely balanced between the two main interests in the nest—the queen and the larvae—a state of affairs aimed at in many a human country's written constitution.

The larvae, as noted, were constantly licked and cleaned as well as being fed, but they passed no feces, which accumulated in a special sac in the lower abdomen. These little black spots would be left behind in the pupal case at the time of pupation.

The nurses gave disproportionate attention to the larger larvae. Obviously, a large grub needs more food than a small one, but the larger creatures got more than their share, especially if food was short as, say, during bad summer weather. This was advantageous from the point of view of the survival of the colony. Workers were its great need, so it was better to feed those nearing pupation than those just starting larval growth. Scarce resources were then being invested in the effort most likely to succeed, because an early addition of workers would in all probability lead to an early renewal of supplies.

The larvae needed mild warmth and high humidity, which is why the nurses moved them around the nest, several times a day if necessary, for optimum living conditions. The sun struck different parts of the heap at different times so that the best position for the larvae varied from hour to hour. But the several thousand larvae in Nest A were not all crowded together in the most favorable part, because the ants liked to have easy access to all the larvae and this instinct acted against the other impulse of putting the young into the best sites. The result was that during the day the larvae tended to be scattered around in the galleries and chambers on the sunny side of the nest. If the night was cold, they would be taken to the central area. The instinctive pattern that had developed meant that though the exact best spot for temperature and humidity was not used for all the young, the spreading out of larvae and pupae was advantageous in the survival stakes. It made a greater number of workers aware of the young and strengthened the bonds between the new and old generations, which in turn ensured a good rate of population growth.

The mechanism, produced by gene selection over long periods of time, was not that of the love between nurse and child so often seen in human affairs but purely a chemically controlled activity. The larvae secreted a substance (soma) that the workers liked, "fixing" them on the drug, so they attended the young to get this reward.

The life of a larva in Nest A was comparatively short in June, about eight days in all, after which it became a pupa. The larval life was a period of intense feeding, for enough food had to be eaten to make the adults, which always emerged from their cocoons fully grown and armed, like Minerva springing from her father's head, and, like the goddess, they grew no more. But the ants, and presumably the goddess, learned from experience and grew in wisdom as they aged.

The adult worker ant needed only a comparatively small maintenance ration for herself. Queens required a lot of food because of their large production of eggs. Adult males had to produce a considerable quantity of sperm, so they ate more than workers, but considerably less than queens. The quality of the food needed respectively by the workers and the sexuals was different, too. Workers could get by with a low-nitrogen diet such as honeydew, whereas queens and males wanted high-protein foods. In short, caviar was not supplied to the general and the best use was made of the resources available.

In addition to removing eggs and placing them in the egg piles, Labora spent a considerable time in cleaning the queen, mostly by licking the body secretions from her, thus ingesting the chemical messengers already mentioned. She also collected food for the queen. This meant that when starting on this task she had to have an empty crop, so on setting out, the first thing she had to do was to empty her own stomach. This was achieved by offering food to an outgoing worker, who usually accepted it and thus came under the influence of the queen's secretions, redoubling her efforts in any particular direction indicated by the chemicals it contained. From her own experience Labora knew where the best foragers came into the nest, the points varying with the time of day. She would go to meet the returning ants and, after the exchange of the usual formalities, would secure supplies from them. Stroking the head of the donor ant with her forelegs accelerated the exchange. With a full crop she then hurried back to the queen's chamber, for she had to avoid being stopped and asked for food as a matter of social intercourse by ants she would meet on the way home. It was like a woman hurrying back with the supper meeting a friend on the way and not wanting to stop to discuss the weather, or offend her, but with the additional difficulty of being unable to explain her haste in words. Her movements had to be nicely judged. If she hurried too much, she would start the kinopsis reaction among her colleagues—the stirring up of excitement to draw attention to some particular event such as the discovery of food or an incipient attack on the nest. If she were too deliberate she would be stopped and asked for food.

At times on these return trips, after having collected food from a returning forager, Labora would seek out the sunny side of the nest and simply rest. It was not entirely wasted time, for it allowed some of the food she had just collected to get into her system and recharge her labial glands, those bodies that secreted the high-protein royal food. To keep the glands active she, like a nursing mother, had to secure a diet higher in protein than that taken by an ordinary forager or nest worker. To this end she became quick in recognizing foragers returning with bits of insects or carrion, which she would secure for herself, her companion royal attendants, and eventually the queen.

In early June, Regina I began laying about three hundred unfertilized eggs a day during the five days from June 2 to 6, eggs that, as noted, would give rise to only male ants. Those ova were carried to the hatching piles in the upper galleries, along with the bulk of fertilized eggs, where they all hatched in the ordinary way. After the first molt the nurse ants recognized the special nature of the male larvae and gave them extra food and labial secretion. Only the standard diet was given to the majority of the females, so that very few fully developed adults capable of becoming queens were produced. By laying unfertilized eggs, Regina I saved a little of her falling supply of sperm, but the main reason for the lack of potential princesses among the larvae was the antifemale hormone being produced by the queen, a condition frequently prevailing in one-queen nests. It was actually the nurses and the queen's attendants who made, or did not make, princesses. To form such a being the extra food had to be given, and given within the first three days from the hatching of the eggs; after that it was too late and whatever food was supplied, the result was a worker ant.

When the larvae were fully grown they had to pupate and were helped in that process by the nurses. The worker pupae here showed a strange divergence; some formed silken cocoons and others were "naked" pupae, in which case the development of the adult took place inside a thin pupal skin. The nurse ants buried the about-to-pupate larvae in soil that allowed the silk spun by the larvae to get some points of attachment. The resulting cocoons were smooth and white—what are popularly (and wrongly) called ants' eggs. Nurse ants dug up the cocoons from time to time to see if the adult within was ready to emerge, which they did fourteen to sixteen days after the start of pupation.

The cocoon or pupal skin splits and the adult pushes out its head, usually assisted by the nurse ants. Worker larvae are, however, sometimes capable of emerging on their own. The newly born ant is pale and weak and needs a few hours for its skin to harden and its limbs to be exercised and cleaned. The young adults then join the nursing staff, allowing some existing nurses to go off to other tasks. So far only workers were being made, greatly forwarding production in the colony.

In June the nest was a hard-working machine geared to the great climax of the year: the emergence of the sexuals and thus the hope of the expansion of that particular colony.

The numbers of the different castes of workers in the nest were slowly changing all the time, according to external circumstances. With the fine summer weather the numbers in all grades increased, but the proportions of the different sects—foragers, sorters, royal attendants, nurses, cleaners, aphid guards, explorers, and so forth—did not remain the same as the colony grew. For instance, it was not necessary to increase the number of royal attendants very much, because the queen needed a staff of only about five hundred workers; she could not lay eggs at more than her standard maximum rate (about 330 per day) however much she was fed, nor did she need more ants to carry away her eggs and clean her. But when more food became available in the outside world, the proportion of ants joining the foraging caste increased considerably. In fact, in June it almost doubled because of the rapid increase in the aphid population. More aphids meant more milkers and more herdsmen to protect the greenflies from predators and parasites. The percentage of sorters and cleaners remained about the same, because the big food increase was mostly honeydew, which diet did not pass through the hands (that is, legs, jaws, and antennae) of the sorters, except for occasional quality control, when a sorter would accept honeydew from an incoming forager, more for her own nourishment than as part of her professional conduct. The sorters still had to check the solids brought in and, though their numbers increased, their proportion in the population of the nest did not.

In the dry days of the month a new category of worker arose—the water carrier—because the humidity in the nest had to be maintained and certain workers collected water, mostly in the early morning, from dew, and dumped it in the drying parts of the nest.

The extra food brought in by all this activity resulted in a considerable June population increase, not so much because the queen laid more eggs (although there was some increase) but because of the greater amount of care bestowed on the eggs and young larvae. Fewer eggs were eaten by larvae, and so more hatched. With abundant food available the workers ate no eggs and more attention was given to the feeding and comfort of the larvae and pupae. The emergence of adults from the pupae became more successful, because more help could be given, owing to the increased number of nurses. In view of the limited communications facilities available (no spoken language, no bee dance, no writing, and only a limited memory), the machine was a remarkably successful one, but, in common with all systems, it had its weak points. One feature not yet dealt with is the matter of guests in the nest. All long-standing (for millions of years in the case of ants) successful systems tend to be exploited by a certain number of freeloaders, and wood ants are no exception. They welcome, tolerate, ignore, or oppose a considerable number of alien animals in their nests.

Many authors have been struck by certain parallels with human behavior, and many a moral lesson has been drawn, but, while the similarities are interesting, the two cases are really very different. Both human and ant seek to exploit other animals. Ant guests secure food in a number of ways. Some are filling an available food niche in the scheme of things, being, like Autolycus and his father, snappers-up of unconsidered trifles. Certain mites and small insects feed on ant feces, secretions from ant bodies that have rubbed off onto gallery walls, bits of dead ants, or dropped pieces of pupal cases, and so on. More adventurous guests steal food in transit, and yet more aggressive intruders attack and eat the larvae and pupae.

The guests thus differ from the human sycophant and parasite who usually attack at one remove, exploiting psychological situations to secure their ends. Obviously, there is a high degree of consciousness in the human approach that is absent in the ant's. But there is no hard-and-fast rule. The human sycophant has eaten many a good dinner at his host's expense, and the wood ant may eagerly welcome and feed the intruder bent on destroying the colony.

It should be noted that many of the guests are not total freeloaders and may do a certain amount of scavenging in the nest, relieving the cleaner-ant cadres of some work. Wheeler described symphiles as "this perplexing assemblage of assassins, scavengers, satellites, guests, commensals, and parasites."

A guest species found in Nest A was the ant-nest silverfish (*Lepisma formicaria*), which lived unharmed in the galleries of the nest. It absorbed the nest smell and was rarely attacked.

A silverfish's body is covered with loose scales, so that if it becomes the object of aggression by a suspicious ant it can usually escape by shedding scales and leaving the ant standing.

Labora unwittingly supplied quite a quantity of food to the silverfish, who secured it quietly and by stealth; the silverfish had a great capacity for gentle, inconspicuous movements. These particular guest insects would wait quietly a little way in from the entrance galleries, where the foragers, returning loaded with honeydew, would regurgitate a drop for the benefit of the internal ants. As the drop was passed from one ant to another a silverfish would run quietly and quickly forward, insert its head beneath the heads of the ants engaged in trophallaxis, and then, raising itself on its front and middle legs, quickly suck up some of the drop of food being passed from one ant to the other. At the same time, the silverfish would tap the heads of both ants with its antennae, the signals leading both donor and recipient to believe all was well. Generally, the ants did not notice the intrusion or the loss. The silverfish also live on rejected materials in the nest, such as the infrabuccal pellets, rotting vegetation, and feces. As they are scavengers, the loss they cause to the colony by stealing honeydew is compensated for, to a slight extent, by their cleaning activities. Moreover, as food grew scarce in the late fall and early winter, the silverfish tended to be attacked more frequently and, being unable continually to regrow the waxy scales on their bodies, they increasingly fell victim to the assaults of ants about to hibernate, and thus created a partial food reserve for the colony.

Parasitic mites fasten themselves on the bodies of nest workers and live on scraps of food from their hosts' mouth parts and the secretion forming on their bodies. They are a considerable drain on their hosts' resources. Labora did not escape their attentions. A mite climbed onto Labora's back as she was returning from carrying an egg to her pile on June 6, and it quickly moved to beneath her mouth. A strange feature of these mites is that they place themselves at different points on an ant's body according to their numbers, in order, as it were, to balance the added weight and the irritation caused. It is to the mites' interest to keep their hosts as active as possible, and the genes have accumulated to produce a species of mite doing just that. For instance, if three mites were on the left side of an ant's head they could partially block an eye, disrupt the signals being received by the left antennae because of their presence around the scape, and put the whole head out of balance, making their host less efficient as a warrior or forager and thus putting the future of their own genes in peril. By distributing themselves over the ant's body, they still reduce host efficiency, but to a lesser extent.

A day later another mite attached itself to Labora, whereupon the first creature moved to the left-hand side of Labora's abdomen and the newcomer took up a balancing position on the right-hand flank. Had a third mite climbed aboard, it would have positioned itself under the ant's head, but fortunately this did not occur. Labora continued to clean herself, but the mites escaped destruction for five days by always moving as their host twisted and turned. On the sixth day another beetle guest (*Decathron stigmosum*) greeted Labora in the usual way—stroking her with antennae and forelegs—and proceeded to rid the ant of the mites. This beetle was a general scavenger in the nest and assisted the colony by keeping it mite-free.

Another guest in Nest A, a Staphylinid beetle known as *Xenodusa cava,* was considerably more dangerous than the mites. In early June an adult female *Xenodusa* followed a returning forager into the nest and was immediately challenged by an outgoing ant. The beetle discharged a repellent spray at the ant, causing her to stop the attack, then hurried forward into a wider gallery in the nest. The creature was a small (one-fourth-inch-long) beetle, broad and dark in color. On its back it had some bunches of hairs known as trichomes, carrying a sugary secretion. The *Xenodusa,* on reaching the wider gallery, stopped on its floor and remained quiet for an hour or so. It then started to advertise its presence by emitting its special fluid which was sweet, narcotic, and very attractive over the trichomes. The *Xenodusa* beetle took care to make its first approach to an ant, a creature that needed food, with a more or less empty crop. One such ant was soon eagerly feeding at the trichomes. The fascinating smell spread around the *Xenodusa* and was soon drawing in a crowd of would-be consumers.

More and more ants became avid for this liquid and constantly courted the beetle, stroking the sides of its head with their forelegs and antennae to get it to emit its fascinating secretion. The *Xenodusa* now found itself welcomed everywhere, for after two days it had acquired the nest smell and was no longer challenged. It followed a nurse ant back to the larval chambers and there fed its fill on the larvae and pupae, while the nurses, intoxicated by the secretion on the trichomes, paid no attention to the devastation being wrought among the young. The beetle then started to lay eggs on a pile of ant eggs in one of the galleries. At once, the nurses set about cleaning and tending these ovae.

Needless to say, the *Xenodusa* larvae, on hatching, fed at first on ant eggs in the pile. After the first molt they were given food by the nurses and buried in the earth when they were ready for pupation; the nurses were deceived throughout into lavishing as much care on the strangers as if they had been offspring of their own queen.

Nest A was now in danger of being destroyed by the *Xenodusa* invasion, for the adults arising from even one female beetle would have produced enough secretion to have corrupted the whole nest; it would have fallen to drug addiction. The ants, eggs, larvae, and pupae would be increasingly neglected as the workers became more and more avid for the secretion. Caring for the young, foraging, sorting the food, and feeding the queen would have stopped and the nest would have died. The beetle secretion also has the effect of producing adults known as pseudogynes, usually having the thorax of a queen with the gaster of a worker; they are not of much use to the colony. Once the *Xenodusa*s had eaten all the ant eggs, larvae, and pupae, they would have walked out of the defunct nest and have searched for new colonies to exploit. So powerful is their drug and so successful are the *Xenodusa*s that it may seem surprising that they have not entirely extinguished the race of ants, and thus themselves, forever.

Xenodusa, or *Lomechusa* (in Europe), addiction in ants is the nearest parallel to similar behavior in humans. Drug addiction can destroy a society[7] in both phyla.

Fortunately for the ants, the nurses were so devoted to their changelings that they treated them in exactly the same way as their own young; in fact, at times they gave precedence to the strangers. For instance, in times of danger the nurses might carry *Xenodusa* larvae and pupae to safety in preference to those of their own species. Paradoxically, similar treatment was the undoing of *Xenodusa.* The ants, it will be remembered, buried their grubs so that they might pupate when the time came, and during pupation they continually dug them up and carried them around the nest to parts they considered to be the most suitable for their best development. They treated the *Xenodusa* pupae in the same way, whereas, in fact, they needed to be handled differently—to be left alone quietly to pupate and emerge from the soil by their own unaided efforts. When hatching time came the ants dug up both kinds of pupae and cut them open to help the young adults emerge. The result of this interference was that not many *Xenodusa* pupae hatched to adults. Yet in another way this difference was also beneficial to the beetles because, if some such restraint had not been imposed, the race of wood ants would have been utterly destroyed and have left *Xenodusa* with no source of livelihood. That is an event the human blackmailer, hijacker, and parasite have to be at pains to avoid, too, if they are to continue to profit from their way of life.

The fact that the *Xenodusa* pupae needed to be left alone to pupate successfully acted in their favor in the later stages of the invasion, because by then the addict nurse ants were neglecting their duties and no longer digging up the pupae, a state of affairs that suited the invaders to perfection. Had the *Xenodusa* needed exactly the same treatment as the ants, they would then have perished from neglect, as did the ants' pupae. This differing need shows how the lives of rival species dovetail into each other, the whole process being arrived at by the push and pull of natural selection over the millennia. The fact is that either some sort of fitting together is reached or one or both species become extinct.

The dangerous beetle laid 150 eggs scattered around the various egg piles in Nest A. A few beetle eggs were eaten by young ant larvae, but most of them hatched to grubs, which fed on the eggs and larvae of the ants (and on one or two late beetle eggs). The beetle grubs were also fed by the nurse ants with honeydew and the usual food supplied by the foragers. In exchange, the grubs emitted small amounts of *Xenodusa* secretion, but it was not as narcotic as that put out by the adults. It had a nice balance, containing enough of the drug to hold the attention of the nurses without being strong enough to make them neglect their duties.

Other strange creatures in the nest were Lycaenidae, the caterpillars of various species of blue butterflies. They impose on the wood ants in much the same way as do *Xenodusa*, but the attractive secretion does not have the same narcotic, poisonous effect. For the first two or three instars the Lycaenidae caterpillars feed on woodland or meadow ground-keeping plants. In the third instar the caterpillar develops a honey gland and, after a molt, falls to the ground and awaits the attention of an ant. In due course it is discovered and the young creature starts to exude honey. On one occasion near Nest A a passing ant tasted the grub, walked around and around it, fed again on the nectar, and drew the attention of other ants to the prize, who seldom tried to deprive the original discoverer of its booty. After a while the caterpillar humped itself up, the ant seized it by the thorax, and carried it into the nest, where its secretion was greatly appreciated. The caterpillar lived on ant larvae and was not attacked by the larval attendants. In due course it pupated, emerged as a butterfly during the summer, and found its way out of the nest to resume the outdoor part of the cycle.

Lycaenidae butterflies have a covering of waxy scales on their wings and bodies; if attacked by ants on their journey from pupal emergence to daylight, they shed the scales and gum up the mouth parts of the ants.

Labora took her share of the Lycaenidae honey but somehow escaped the temptation of the *Xenodusa* secretion and so continued as a useful member of the colony.

In spite of the losses of food and larvae because of feeding so many guests (not all are mentioned here; the full list might be tedious) and the dangers of the *Xenodusa* invasion, Nest A was flourishing. So much food was being collected that the siphoning off of a small percentage to "underdeveloped species" made very little difference to the ant economy. The ants could afford it.

Toward the end of the month the workers were preparing for the great event of the year: the emergence of a flock of mostly male ants. In their timing, wood ants were different from most other ant species, where the emergence of the sexuals is at the end of summer—the fat season—when food is abundant. Wood-ant efficiency is such that they can have a comparatively early sexual brood, giving the new queens a better chance of establishing fresh colonies during the late summer.

Great care had somehow been taken to slow down the development of the early-laid eggs (June 2 and 3) and hasten the emergence from pupation of the later eggs (those laid June 4, 5, and 6), so that the sexuals would fly out all together. A light rain shower on June 26 had made conditions right, and the worker ants set about enlarging the exit galleries and the main exits from the nest so that the comparatively large-winged males, and a few females, too, could get out. The chemical messengers circulated through the nest and created an atmosphere of general excitement. Labora was still one of Regina I's attendants and was kept busy following the queen around, because all this activity in the nest had disturbed her placid life of feeding and egg laying. In Nest A Regina I was the only source of the queen-substance chemical, and the ants acted as though they were aware of that and had to retain her. Regina I, on the other hand, was controlled by her genes to favor their propagation to the utmost. To do this she needed sperm, and, from the various internal pressures in her body, she knew that her supply of it was dangerously low. If she ran out there would be an end to the direct propagation of Regina I genes. In other, and simpler, words, an urge came over her to mate again, a desire induced by the emptying of her seminal vesicle.

Much the same state of activity prevailed in Nest G, where the sexuals, nearly all fully developed females, were being prepared for emergence. Nest G, it will be recalled, was a multiqueen colony, and the hormone control on the production of princesses had been such that they all emerged from their pupae about the same time. In both nests the sexuals were controlled by the workers, by being kept inside until the external conditions were right—warm and humid. This gave an opportunity for their wings to harden, their limbs to be exercised, and a food reserve to be accumulated against the ardors of mating. The adult sexuals, particularly the female ones, were fed mostly on honeydew, which was still plentiful, but flesh (mainly small insects) was also given to them from time to time. Strangely enough, the males within Nest A did not try to mate with the few princesses—their sisters—inside the colony; they needed open-air conditions.

6

JULY

AT 11:00 A.M. on July 1 the winged males began to emerge from Nest A, and the winged females from Nest G and some other nests in the neighborhood. Worker ants were at all the exits in both nests to help the sexuals get out and to prevent any queens from emerging. In Nest A, Labora and a number of her companions followed in the wake of Regina I who, as mentioned, seemed most eager to get into the daylight. She actually got her head outside Exit 14 on one occasion, but Labora was able to seize a hind leg, brace herself against a gallery wall, and hold her royal mistress back. The little worker emitted the alarm substance; other ants hurried to her help, and three of them dragged their queen inside the nest, closed the gallery, and offered her food. The queen accepted some labial secretion, giving Labora time to seal another upward-sloping gallery leading to Exit 12. By skillful manipulation and the blocking of galleries by the bodies of several worker ants, the queen was persuaded to return to her egg-laying chamber and was quiet for a little while. Nevertheless, the royal attendants had to keep a constant watch on her.

About a thousand males from Nest A gradually accumulated in a swarm over a patch of bare ground at the edge of the pine wood, flying backward and forward, around and around, and occasionally resting on the leaves of bushes or tall plants. They were there for three days. Meanwhile, the princesses from Nest G were steadily emerging but did not gather in such large swarms as did the males.

At times, one of the males would leave the swarm and dart a few feet away. Unless he found a princess, he would soon rejoin his companions. It is a curious feature of the life of wood ants that single-queen nests tend to produce the males, whereas female sexuals dominate in multiqueen colonies.

Occasionally, a female from Nest G would fly toward the male swarm, and one ant there would pounce on her, either in the air or when she had landed on the bare patch or a leaf. The males had more facets in their eyes than workers or princesses and were very alert to the movement of females. There was very little courting behavior, and usually mating took place quickly. Most of the females only mated once, and some of them bit off the gaster of the male toward the end of the process. The mutilated ant would then let go, crawl away, and slowly die. Other more fortunate males escaped and flew away in search of food. They usually did not seek to mate again, as nearly all their sperm had been deposited in the first female; a second mating would have had very little biological meaning. They lived a few days on their own and slowly died. Solitary ants of any sort do not seem to be capable of much intelligent effort; they need the stimulation of other ants around them in order to undertake their various tasks.

The swarms of males, females, and copulating couples were attacked by a number of predators—particularly birds, other ants, toads, and frogs—but large numbers of wood-ant princesses, now mated, survived.

The next task of such a female is to secure her future by founding a nest or securing an entrance to an existing nest, of which the last possibility has several aspects. Wood-ant princesses are rarely successful in starting a new colony entirely on their own, for it means finding a site, driving a tunnel, laying eggs, foraging for food, and tending the eggs, larvae, and pupae until the first worker offspring emerge, after which the queen reverts to being an egg-laying machine and the workers feed her.

Most of the mated princesses that survived gained entrance into their own nest, into a sister nest in a loosely knit colony of nests of the same species, or into nests of another species, of which the last created a strange situation, to be dealt with in a moment.

The mated females, on landing, cast off their wings by rubbing them against their legs, projecting stones and earth. Two purposes are served by this loss: first, it becomes easier to get into and move around in the nest, and second, the substantial wing muscles dissolve and are absorbed into the system, forming a considerable food reserve for difficult times ahead.

Some of the mated females from Nest G landed near Nest A and tried to gain access to it, but the foragers and guardians there had become so imbued with their own nest's smell and so conscious of their need to retain their queen—the source of the future existence of their colony—that they rejected this easy solution to their problem. It was not that blind loyalty to an old lady who had served them well for years and was now, regrettably, failing, but that particular circumstances had rendered the nest incapable of recognizing the benefits of a change. But not all the advantages were necessarily on the side of change. The independent observer might note that it was perhaps better to stick to an old and tried source for eggs than to adopt a new one of unknown genetic composition. Suppose, for instance, that a new queen carried genes inducing, say, an instant welcome to any *Xenodusa* seeking entrance to the nest, or genes making the nurse ants leave *Xenodusa* pupae alone in the soil (instead of lavishing ant pupal care on them, which did the beetle pupae so much harm); then the colony would have been doomed. The gene bank is a lottery with but few prizes, especially for the ants whose genetic changes are very slow. Nevertheless, the possibility also exists that a new queen could introduce advantageous changes, too, such as better navigation, more efficient use of fewer sexuals, and the introduction of a language of definite antennal tapping, or even the bee dance, leading to the solution of many communications difficulties. But it is a little presumptuous for man to lay down the law on what would, or would not, suit ants. They are a very much older civilization than ours.

When a mated Nest G princess, after throwing off her wings,

approached an entrance to Nest A, she was at once set on by the two or three guardians of that particular door. It was a question of timing; as the workers attacked, the would-be queen made conciliatory gestures with her forelegs and antennae and regurgitated a drop of food rich in queen pheromones, but she was not quick enough to overcome the initial resistance of the Nest A workers, who, having once started on an aggressive course of action against that alien smell, were conditioned to continuing it. The princess was pulled back by her two hind legs by two workers and another then seized her body. One of the first workers stretched and twisted her leg, and a fourth worker (for the excitement had created a considerable body of opponents) reared up, twisted her gaster around, and sprayed the princess's leg joints with formic acid. The skin there, as mentioned, is thin and the acid effectively paralyzed the potential savior of the colony. The workers then dragged her down the ant heap to the rubbish area and left her, where she recovered, moved off, and attempted once more to secure her future. She slowly traveled back to Nest G, occasionally exchanging food with G workers and strengthening her G smell, and was eventually admitted there.

When Regina XVI (the princess we have been considering, whose admission to a nest can be considered the equivalent of a coronation, entitling her to the name Regina) got inside her nest, she immediately brought up a drop of food into her mouth and offered it to the entrance sentinel. As the newcomer already had the general smell of Nest G, there was no difficulty on that score; nevertheless, the workers could have rejected the advances of a new queen if the existing queens had been secreting enough of the antifemale chemicals. This was not the case; the sentinel at once took the proffered drop, stroking the new queen's face with her forelegs. The two ants then set off toward one of the upper chambers containing only a few workers. Here, the new arrival moved around tapping antennae and offering food in the absolute dark. The antennae indicated that the smells were right, as were the shapes. The small amounts of queen chemicals found in the regurgitated food attracted the foragers, who happened to be in that chamber, to her service and soon they were cleaning her with their mouths. The secretions Regina XVI sweated out over her skin contained far more of the attractants and queen substances than did her stomach, and the body secretions turned the foragers from casually and temporarily attracted acquaintances into firm members of the new royal household, so to speak. In short, she now had a couple of dozen workers in her service, enabling her to get a firm foothold in her new territory. The other queens in the nest were not antagonistic, and soon Regina XVI was laying eggs, the workers carrying them indiscriminately to any of the many egg piles in the colony. Only fertile eggs were laid and no special feeding was given to the larvae, so only workers were produced.

The number of mated princesses produced by a colony the size of Nest G can be very large, say about two thousand, and all cannot be accommodated in the terrain available. About half of them fall victim to predators, still leaving a large number to find homes. It might be asked why, since all the princesses carry the home nest smell, they do not all reenter that nest. The answer is that the number of queens in the nest is nicely regulated by the resources of the colony, mainly the food available and the size of the nest. If too many princesses are taken in, more and more anti-female hormone will be produced, and, if this substance is not balanced by the availability of more food, then the workers will react to the chemical signal by refusing entry to more females and may even kill newly established queens. It is a form of birth control, the population being regulated to fit the resources available, reminiscent of the ancient Incas, who fitted population to food resources by confining surplus females in convents until past the age of childbearing.

The strangest way the young mated females have of establishing a new colony is to take over an already existing nest of another species of ant. The *Formica fusca* group often gives them an opportunity of doing this, especially in the case of small nests. The mated *rufa* approaches a *fusca* nest and tries to ingratiate herself with the emerging foragers by offering food. Frequently, in the neighborhood of such a nest she has to fight *fusca* workers. Being bigger she usually wins, and, in tumbling around with the alien ants, she absorbs a little of their smell. The *rufa* princess can then approach the *fusca* nest and attempt to enter it, using the usual conciliatory technique—the offer of food. With some slight hesitation about the smell, the *fusca* guards may pause long enough to have their doubts overridden by the offer of good nourishment. (How many human sentries on a cold night have been suborned by a generous dram of whisky?) Once the guards have taken food from the intruder she is on the way to acceptance; more workers meet her and exchange food, and finally she takes up residence in an empty chamber. After about a week, when she is thoroughly acclimatized to the new nest and knows her way around, the *rufa* princess makes her way to the *fusca* queen's room and bites her head off. The workers remove the dead queen and adopt the new one, who starts active egg laying.

The eggs, of course, are *rufa*s. They are carried away by the unaware *fusca* workers to the egg piles, cared for in the usual way, and eventually hatch, become larvae, then pupate and emerge as adults—*rufa* adults—who mix and work with the *fusca* owners. The *fusca* workers gradually die off as they age; the nest then becomes a pure *rufa* colony.

Wood ants have thus transferred much of the trouble and difficulty of colony foundation to another kind of insect. Nevertheless, such behavior carries a risk. To have a nest built for you is a great saving of effort and, in the comparatively short term, genes might well accumulate that lead the wood ants to rely entirely on *fusca* for new nests. If at the same time the build-your-own-nest genes were lost, *rufa* would then become dependent on *fusca* for their long-term existence. Obviously, if all new *fusca* nests were always taken over by *rufa*, eventually *fusca* and finally *rufa* would vanish as species. The wood ants might be saved by some build-your-own-nest genes still around or even arising again by mutation; if these came to the fore, the old habit could be reestablished. If not, it would be a question of balance—of sufficient *fusca* nests being left alone to support both *fusca* and *rufa*.

As things are now sufficient, *fusca* nests are left to ensure the continuation of both species, conforming to the usual parasite/host in nature; the exploiters leave enough of the exploited alive to ensure the continuation of both groups. The wood ants also extend their sphere of influence by expanding multiqueen nests and forming loosely connected daughter colonies. It is even possible that in an extremity a *rufa* princess could establish a colony on her own. If she could also induce a few workers to help her, matters would be that much easier.

The mated *fusca* females are never readmitted to their own nests and thus have a built-in propensity to establish new colonies, which they do on their own comparatively easily. The new queen digs a cavity under a suitable stone or in a rotten tree stump, lays eggs, tends them, hunts for food for the larvae, watches over their pupation, and sets the resulting workers to attend her. The nests are usually little ones and no mound is built. The fact that the nests are numerous and small means there are greater opportunities for *rufa* princesses to get started by capturing a nest. Of course, when *rufa* takes over such a nest the character of the nest changes and a mound is started, which after ten or twenty years can be of considerable size.

The takeover of a *fusca* nest by a *rufa* princess is not always a smooth enterprise. Even after the young queen has gained entry, she may not acquire the nest smell sufficiently rapidly and thus may be attacked on one of her journeys by suspicious guards. The would-be new queen may also set out on her Lady Macbeth enterprise too soon and find that the resident queen she has set out to murder resists the attack and is perhaps aided by her staff. Nests have been found taken over by young *rufa* queens with, in one case, at least four dead *rufa* royals with her, killed by the resident queen and her attendants. It is even possible that a *rufa* princess, having established herself and killed the *fusca* queen, then sets about killing any other *rufa* females that try to get into her chosen area.

An interesting feature of *fusca* nests is that among a variety of guests some lightning bugs (Lampyridae) may be found, feeding, as is usual with myrmecophile beetles, on the brood. Although it is the female that emits the most light, the males, eggs, and larvae are faintly luminous. It is difficult to say whether the depredations caused by the Lampyridae are compensated for by the provision of light in the nest. On the whole, ants seem to get on very well in the dark, since their senses of touch and smell are so keen, but the advantage of being able to employ sight may well be worth the loss of a few larvae, precious though they be.

The fact that *Formica fusca* is the ant found in Baltic amber, laid down in the Eocene 50 million years ago, suggests that this insect has not suffered too much from the social parasitism of *rufa*. After a long time the species is still going strong.

Four small *fusca* nests (B, C, D, and E) in the clearing in the Adirondacks, near Thirteenth Lake, were entered by princesses from Nest G. In a year's time the nests stood every chance of becoming *rufa* colonies. Six princesses were readmitted to Nest G, and four of them were able to set up their own egg-laying center with a team of attendants; the other two disappeared. This means that out of the three thousand sexuals produced, only eight females had any future. They had been mated by eight males, who either had been killed in the act or had died shortly afterward. On the face of it, the waste of life seems enormous—just under 99½ percent of the sexual population produced was unused—but in fact the increase of eight queens on the fifteen in the two nests was very large, and such advances could not have been supported for long without turning the area into a seething mass of wood ants. It is a characteristic of insects, and particularly of ants, to waste their seed, because the chances of any one female being successful are very remote. But if it does come about that circumstances are favorable to the species, then there are plenty of individuals waiting to take advantage of them. A swarm of locusts is an example. This is merely to say that insects can multiply rapidly—a matter seen strikingly in pest-control operations. An insecticide that kills 98 percent of a pest population might be considered to be almost as good as one killing 99 percent, but if the pest (such as aphids) is capable of rapid reproduction from the survivors, the former spray is only half as good; it leaves 2 percent of the population behind to breed up, whereas the second treatment leaves but half of the former chemical's nucleus.

Let us consider the fifteen queens in the two nests; each one had an average life of ten years, and they had been added one by one as the nests grew. Queens would die (or be thrown out because they had stopped laying eggs or had run out of sperm) at the rate of about three in two years, or one and a half per year, the figure needed to keep the population steady. The addition of eight new queens in this particular year presaged a vast population increase of over 500 percent. Such a rise under normal circumstances would have been impossible and would not have taken place.

A moderate population increase was possible for the two nests, and the addition of the eight extra queens did not necessarily mean that that many more adults would be hatched, for two reasons. First, the new queens would not all be well fed, and second, their rate of egg laying would drop off. More important still, nothing like so many of their eggs would hatch. Because of an insufficiency of food to support all the new brood fully, a considerable number of the new eggs would be eaten by the workers—the nurses and foragers. In addition, a certain amount of automatic population control took place in the egg piles. If the larvae were neglected after the first molt, they would continue to feed on the eggs, thus restoring the population balance. Dangerous guests in the nest would be less opposed and challenged, and they consumed eggs and larvae, too. In fact, the whole program of worker births would depend on the amount of food available. If it became plentiful, because of good weather and lack of enemies among the aphids, the new queens were there and their eggs would then be used for hatching rather than for consumption. The situation is closely analogous to a poultry farm turning from the production of eggs for eating to eggs for breeding, in response to a sudden demand.

Moreover, the "unused" sexuals were not wasted in the ecological sense. They became food for a considerable number of other animals such as birds, toads, lizards, insects, and other ants, including some wood ants.

Toward the end of July the excitement in the two nests was

dying down. Regina I did not have to be watched quite so closely and had settled back into a routine of laying fertilized eggs, but at a rate rather slower than before. Nevertheless, the workers coming forward were all kept busy on nursing duties, and the former nurses began new tasks, as was customary, except that with things now so easy, life allowed them a little more leisure. Sunbathing, strolling around with the gang, and mock battles could be permitted to take up more time. Meanwhile, the *Xenodusa* larvae were starting to pupate—a time bomb in the entrails of Nest A.

In Nest G things were different: four new queens to be cared for meant a lot of additional work. The biggest call on the work force was for considerably more labial secretion, to keep the new queens in first-class form, as well as to supply the ordinary run of food. Quarters had to be prepared and enlarged, and even if most of the eggs were eaten they still had to be carted away and the new queens cleaned. At the end of the month Regina IV died and her staff dispersed; some of them with plenty of labial secretion available attached themselves to one of the new queens, Regina XVI.

7

AUGUST

XENODUSA was mentioned at the end of the last chapter. At the beginning of August, a little late, the nurse ants were digging up the beetle pupae and moving them, along with naked and enclosed worker-ant pupae, to spots in the nest that the ants considered favorable for emergence. Such spots, as noted before, did not suit the *Xenodusa* at all; they needed to be left alone. As a result of this misapplied effort, only five adult *Xenodusa* emerged out of about a hundred larvae that actually reached pupation. However, five adults (three males and two females) were a serious threat to the existence of Nest A; they could have corrupted the whole colony.

Fortunately for the ants, at this comparatively late season of the year the *Xenodusa* secretion had lost some of its power and the Nest A ants had grown more suspicious. The first male to hatch (X1) arose from a pupa that had been overlooked by the ants, as it had become interred at some depth in a gallery wall. As a result, X1 did not have a strong smell of the nest. After emerging from his pupal case he bored his way out of the soil, and before his wings were dry or his limbs hardened, he was challenged by Labora and two other workers who happened to be moving along that roadway. X1 was not yet ready for such an attack; he made an attempt to shoot his repellant spray at his attackers but hit only one with a very small drop of it and was quickly seized by Labora and one of the workers. Labora twisted her abdomen between her hind legs and shot a spray of formic acid forward and upward into the beetle's face. The creature was halted, became unconscious, and was cut up and handed over to the sorters and scavengers. The edible parts were retained and the waste was carried out to the dump heap, the border of detritus that lies all around the base of a wood-ant mound.

The two other *Xenodusa* males managed to emerge successfully from their pupal cases, in spite of the help being lavished on them by the nurse ants, and moved toward an exit in order that their limbs might harden and dry. They were attracted outside the nest by the sunlight, and, after half an hour, they were in first-class condition and ready to reenter the nest. The guards would not let them in, even though the beetles tried to make themselves attractive by ejecting the fatal fluid over their trichomes. Had the ants been human, a splendid moral lesson could have been drawn—they had signed the pledge, promising "never to touch another drop." Wheeler pointed out in 1925 that the trichomes diffuse an aromatic secretion having the attraction of a good cigar to man, or catnip to a cat. To continue the human comparison, in Nest A it was now as if the *Xenodusa* secretion had only the power of tobacco over its (human) addicts rather than that of cocaine; not that the former cannot kill just as effectively, but it does not destroy the will, as do the hard drugs. It is a matter of timing; the "pushers," beetle or human, have got to make their addicts quickly if they are going to succeed. By early August, after the mating flight had taken place, the wood ants did not succumb so quickly to the charms of *Xenodusa*, rather like the small boy who, having seen his elder brother make himself sick on one of his father's cigars, eschews the weed for some time to come. Previously, the ants took one sip and were hooked; now they paused. The fact is that late in the season a *Xenodusa* instinctively knows that it is no longer persona grata in wood-ant nests and moves off to those of another species of ant (usually a *Camponatus*) in order to pass the rest of the year. This is another aspect preventing the total extinction of wood ants by beetles. *Xenodusa* moves out voluntarily before too much harm has been done.

One of the female *Xenodusa,* X4, did attract a customer for her secretion shortly after her emergence from the pupal case, but being aware of the need to find new quarters, she seemed to regard the worker ant more as a hindrance than a potential supplier of food and shelter. As soon as the worker had taken its fill, X4 moved unobtrusively to an exit gallery, left the nest, crawled down the ant heap, and started searching for a *Camponatus* nest. She was eaten by a ground shrew before she found one.

The fifth adult *Xenodusa* in Nest A, X5, moved out of the nest immediately upon emergence, being hastened on her way by some aggressive guard ants. She did find her way to a *Camponatus* nest (Nest F) in the neighborhood, mustered her secretion, obtained entrance, and found a niche in a deep gallery, which she left from time to time to feed on the ant larvae. She was unobtrusive and quiet, did not have any more occasions to use her secretion, and devoted herself to preparing for the hibernation to come. The two male beetles also found the *Camponatus* nest and took up their winter quarters in it.

Ten yards east of Nest A was a colony (Nest H) of another species of ant, the blood red robber ant (*Formica sanguinea* group), famous for being just that. The nest was situated under a stone at the edge of the pine wood and was about two feet in diameter. There was no mound and the galleries were all underground. To protect it from excessive summer heat the ants had covered it with a carpet of dried grass and broken pine needles about two inches thick. The covering blew away in winter, allowing the nest to absorb heat from any winter sun available. It was a large nest and thus an aggressive one, which was bound to clash sooner or later with the other ants' nests in the neighborhood.

Robber ants mostly raid the unfortunate *fusca* nests, but, as noted, they will also attack wood ants and other genera, too. They capture eggs, larvae, and pupae and carry them back to their own nests to be used as food or to be raised there, to emerge as adults and in due course wait on the *sanguinea* as attendants. Such relationships between ant species are usually referred to as slavery, but a few myrmecologists think this term should not be used of this ant association. As the situation is so like slavery, as it presents interesting parallels and differences in regard to human behavior, and as the word was used by Wheeler, the expression will also be used here.

First, as to differences, this slavery is without the immorality of human subjugation. Second, the enslaved ants willingly work for the greater success of their mistresses and of antdom in general. Third, they leave no offspring, because queens are not enslaved; and fourth, the slaves are not aware that they are slaves. Thus, the "antislavery" myrmecologists could say that the situation was one of cooperation between two species, but this is not the case, because the slave genes are disadvantaged in the struggle for existence—they leave no descendants.

Some species of ants rely very much on slaves; in fact, *Polyergus rufescens* is utterly dependent on them. *Polyergus* ants cannot feed themselves and rely on a slave species (once more, *fusca*) to do it for them. *Fusca* puts the food into their mouths.

For instance, P. Huber put thirty *Polyergus rufescens* into a chamber with plenty of food and some larvae and pupae to stimulate them to work. They were unable to fend for themselves or even to feed the larvae, and many of the "master race" died of starvation. Huber then introduced one *fusca* worker, who instantly set to work. She fed and saved the surviving *Polyergus*, made cells, and cared for the young.

However, the robber ants do not go to the extremes that *Polyergus rufescens* do. They can live perfectly well without their domestics, so that if they destroy all the *fusca* nests in an area (which can and does occur) they are not totally lost, as would be *rufescens* without *fusca*. However, *sanguinea*'s *fusca* and wood-ant slaves are not quite a luxury either. Although their presence does make life easier, the slavery is more a device to increase the survival efficiency of the *Formica sanguinea* group, and consequently the attendant existence of *fusca*. An interesting parallel with man thus exists. Slave owners always maintained that their slaves—"backward" peoples—were far better off under their paternal care than if they were set free to face the world unaided, a view also put forward by the African chiefs who sold the slaves in the first place.

Formica fusca—the big black ant—seems fated to be a lowly but necessary wheel in the grand design of the ant machine, always suffering, but important to many species.

In the world, things are very often not what they seem to be. For instance, when the robber-ant colony makes a raid, it often seems as if a leader determines on the action, sends out scouts to reconnoiter the terrain, launches an assault, and sends for reinforcements in order to follow up an advantage or remedy a repulse. The ants then mount flanking attacks, surround the target, storm it, and emerge victorious. That is what appears to be happening, but it is a sequence seen in terms of human aggression. What actually happens may be seen by following one of Nest H's slave raids.

Ants from the *sanguinea* nest (H) were out early on the morning of August 4. A group of activator ants were soon climbing the slopes of Nest A, much to the consternation of the wood-ant workers there, who repulsed them one by one, killing a number in the process. The robber ants then returned to their nest in an excited state and, by means of the kinopsis reaction, gradually got their colleagues to turn out and walk approximately in the direction of Nest A, in ones and twos or in small groups. They gave the impression (as was probably a true one, given the ants' communication difficulties) that they had been advised that something advantageous was afoot but they did not quite know what. A column of ants about nine feet wide advanced slowly in that direction, with only about six of them to the square foot. If a member of the expedition found food, it would usually stop to examine it and then turn around and carry the morsel back to the nest. It was naturally the ants leading the column who tended to discover the food, and, consequently, these leaders were frequently passed by ants from the center of the column, while in due course the center ants were overtaken by the rearward ones.

One of the small *fusca* nests (D), which had been taken over by a wood-ant princess from Nest G, lay in the path of the robber-ant expedition and was discovered by a solitary worker. Although excited at discovering a prize, the lone ant appeared to hesitate as to the advisability of making an attack on its own. However, a single ant by itself is usually a most unenterprising creature, so this particular robber merely retreated from Nest D, went back a bit and around the nest, and continued forward. Although it did not even try to spread the news of the discovery but merely ignored it, the ant gave the appearance of having sent back for reinforcements, for a few moments later, six robber ants in a loose group came across this same nest and attacked it. Fights developed between the *fusca* workers on the south side and five of the robbers. The defenders started to call for help by spraying out their alarm substance, thus leading the robber ants to bring chemical warfare to support the attack. They started to spray out what F. E. Regnier and E. O. Wilson call the propaganda substance, chemicals derived from the *sanguinea* workers' much-enlarged Dufour glands. As much as 700 micrograms of this substance—a mixture of various doceyl acetates—was to be found in each worker's Dufour gland. These volatile acetates have higher molecular weights than the usual alarm substances found among ants, with the result that the propaganda chemicals evaporate more slowly and remain active for a longer time than the ordinary alarm signals. The propaganda substance frightened and dispersed the *fusca* workers but had quite a different action on the *sanguinea;* the robber ants themselves were only excited by, and attracted to, the chemicals.

More than a hundred years ago, Huber noted that *fusca* seemed cowed by a *sanguinea* attack. Indeed, so frightened were they that they often would not reenter their own nests after the robber ants had taken their booty and retired. This is now known to be because of the persistence of the propaganda substance and the fact that it completely disrupts the chemical signals used by *fusca* ants in the ordinary course of their lives. The dispersing effect of the spray alone not only makes joint action ("You hold him and I'll hit him") by several ants less likely but also lowers the abilities of solitary individual ants. Ants need companions in order to be stimulated to effective work.

Not all the defending ants had been affected by the propaganda spray, and some *fusca*s started carrying out their pupae on the opposite side of the nest; about a dozen were got clear, but by now more robber ants had arrived and had surrounded the *fusca* nest. That was not a deliberate flanking movement or siege tactic but the result of each robber ant acting on her own; each one sought a hitherto unoccupied point of attack, and thus the target became circumscribed.

The escaping *fusca* workers, carrying pupae, were stopped when they reached the ring of robber ants, and a struggle for possession of the precious burdens ensued, which nearly always resulted in a robber victory. The attackers then carried the pupae back into the *fusca* nest, left them in a gallery near the surface, and came out to supervise the escaping *fusca*s once more. The few *fusca*s that managed to break through the robber-ant ring with pupae joined the pathetic band of their companions, some carrying their precious salvage up grass stalks in the hope that once there they would be safe. The ironical thing was that many of the new pupae were wood ants, hatched from eggs laid by the invading *rufa* princess. The *fusca* royal staff now hurried their new queen out of the nest on the north side. The robber ants attacked and killed her with most of her attendants, though they let the majority of the *fusca* workers go once they had removed any eggs, larvae, and pupae they were carrying.

The *fusca*s removed not only the brood but also many of the nest's guests, for they all had the nest smell and had ingratiated themselves one way or another with their hosts and consequently appeared to be of equal value to the brood. Had the dangerous *Xenodusa* been present, equally energetic steps to save them would have been taken, as with their own pupae.

Robber ants are bold and aggressive. *Fusca,* subjected to the propaganda mixture, is weak and submissive; it soon gives up when attacked. It is interesting to note that the pacifist ant has survived as a species for 50 million years. Of course, we do not know if it was always pacifist or has only taken to that course of action over the last million years or so. Yet in spite of its meekness (perhaps even because of it) it survives as a numerous and important species; there are more *fusca* ants around than the fierce *sanguinea*. "Blessed are the meek, for they shall inherit the earth." (Matt. 5:3) The rule seems to apply to a greater extent in the world of ants than, at present, in that of man.

After an hour the *fusca* nest had been cleared of its owners. The robber ants now went in and started removing the booty, making little piles of pupae on the ground outside the nest. Naturally, the piles were a target for birds, shrews, and lizards. However, the majority of the pupae (and a few larvae, too) were carried back to the robber-ants' nest. About half of them were eaten and the rest put into brood chambers, where they would hatch and be incorporated into the robber-ant economy. But they did not entirely lose their *fusca* characteristics; they were perfectly prepared to help their mistresses in slave raids and to capture more *fusca* pupae, yet they tended to build in a *fusca* style when put to that task. Of course, a lot of the pupae captured by the robbers were in fact wood ants, and they, too, would submissively work for the robbers but build in their own fashion.

When the robber ants had cleared Nest D of all its pupae, larvae, and eggs, they abandoned it. One of the few *fusca* workers in the little group that had survived started scouting around for food as dusk fell. She noticed that her own nest was empty and that it was being entered by guest beetles that had been carried out by *fusca* workers or turned out by the robbers, but the frightening smell of the *sanguinea* spray still hung around the place, overriding the nest smell well known to the scout; it effectively prevented her from entering her old home. She hunted around the neighborhood and found a small hollow under a stone, a cavity free from the *sanguinea* gas. The solitary ant returned to the *fusca* survivors on their little hillock and started the kinopsis agitation, running around the workers and excitedly tapping antennae. She then picked up one of the pupae and started to drag it to the hollow. Another *fusca* worker then went to help her, and gradually the rest of the ants got the message. The pupae were all carried to the temporary refuge, small though it was, even for the remnant of the colony.

The next day, scouting was resumed and the old nest was rediscovered; as the *sanguinea* odor had now evaporated, the *fusca* scout recognized the nest smell, entered and explored the nest, then returned to the temporary home. There she once more started the kinopsis agitation and induced the survivors to reenter their old nest. It was too big for them and the colony was queenless. As the ants were now preparing for winter, the queenlessness would not be of much importance, but a queenless nest would not survive the following summer. The colony then would either die out or be taken over by another species such as the wood ants.

Only comparatively small numbers of the robber ants from Nest H were concerned with the successful attack on the *fusca* colony (Nest D). The main front of the broad column moved forward in its characteristic way—the center overtaking the front, and the rear the center—until they reached the wood-ant nest A, the one with which this book is mostly concerned. An attack on a pugnacious *rufa* is quite another matter than a raid on the gentle *fusca.* In the first place, the nest itself is a much more imposing edifice, a great skyscraper fortress; and second, the species has powerful jaws and is courageous.

The *sanguinea* (robber ants) paused in front of this obstacle and, as more and more of them arrived, spread out and around the mound until it was completely surrounded.

By chance, Labora was on the lower half of the south slope of her nest when the robber ants made their first advance up the mound. She and a colleague seized the leading attacker, turned her over, and cut off her head. The still-struggling body was carried into the nest and handed over to the sorters, who quickly cut it up, setting out the soft parts as food. Labora was emitting the alarm substance from glands in her head and gaster, and soon the wood-ant workers were swarming over the surface of the mound, standing up to the invaders in a much more soldierly way than was the habit of *fusca,* who were somewhat cowed by the very presence, let alone ferocity, of *sanguinea.* Hundreds of combats were now taking place all over the mound. As a precaution, the wood-ant nurses were bringing larvae out of the galleries and piling them up on the top of the nest. Formic acid was being sprayed everywhere, enough to deter the woodpeckers and flickers, always on the lookout for cocoons, so that the pile was safe from them.

Labora, as mentioned before, was an activator ant, so she soon returned to the battle, leading the agitation to make a breakthrough on the north side of the robber-ant circle. The struggle was long and furious, but a passage was opened and the wood ants started to carry larvae, pupae, and guests, including one of the dangerous *Xenodusa* beetles, away a little distance.

On the south side some robber ants managed to enter the nest and were bringing out pupae. Within the nest this penetration was resisted and contained, because the wood ants were blocking passages leading further into the nest and removing pupae and larvae deep inside. The eggs had to be abandoned. Having blocked off the *sanguinea* enclave on the inside, the defenders next cut off the access of other robbers trying to follow up the advantage gained. Very few reinforcements got into the *rufa* nest. The next step was to isolate and then destroy the robber-ant foothold. This was fairly easily done, because all that the wood ants had to do was to wait at the exits. As the invader ant came out, usually carrying a pupa, she was attacked by two or three of the home ants, the pupa rescued, and the kidnapper killed. The pupa was then carried back into a safe part of the nest.

Naturally, all the inhabitants of Nest A became excited by the attack, and the general agitation once more disturbed Regina I. She stopped laying eggs, moved out of the royal chamber, and wandered around the nest. The guards could not pay a lot of attention to her because of the confusion created by the *sanguinea* attack. Although it was late in the year, Regina I was once again seeking a male in order to replenish her sperm store. This powerful impulse overrode the alarm signals, and she ignored the clouds of formic acid everywhere as well. She was determined, somehow, to get out of the nest.

Labora found Regina I working on enlarging an exit gallery, obviously in order to escape. She seized the queen, dragged her back, and was helped in the struggle by two other workers passing by. They all joined in the contest to restrain their queen, the source of their most precious possessions, their only wealth—the young. Regina I fought energetically to free herself, her motive being the same as that of the workers—the need to save the colony, to do which she needed sperm in order to lay fertile female-producing eggs—but her efforts were to no avail. She had lost one of her main weapons in wandering around the nest; much of the queen substances had rubbed off her as she went along the narrow galleries, and more would be secreted only when she was at rest and being fed, so she could not offer the royal exudation that might have quieted her attackers. In exchange she had learned much about the layout of the nest and now could quickly find her way about it. Regina I was big and did not give up the struggle easily. She bit off a leg from two of the attacking ants and incapacitated them to some extent. Labora then resorted to desperate measures. While the two wounded ants were holding Regina I down, Labora approached the royal head, reared up and twisted her abdomen forward between her hind legs, and slowly produced a drop of formic acid from the end of her gaster. Holding the queen's head with her forelegs, and helped by the two ants, Labora felt with her antennae and palps and, locating the queen's mouth carefully, inserted the drop of acid into it. The effect was startling. The queen was knocked unconscious. Labora and some other ants who had joined in the fray carried her back to her quarters.

The wounded wood ants were still reasonably active, managing to get along on five legs instead of six, and they had helped move the unconscious queen. They and Labora left her in the royal chamber and returned to the scene of the robber-ant invasion, intent on destroying the invaders and rescuing the hostages (the pupae). Getting along on five legs is one thing, but fighting a robber ant with that handicap is another. The two wounded ants were both killed in the struggles. Labora was very busy, rescuing a considerable number of pupae.

After an hour Regina I recovered consciousness and her full faculties. She set off once more, choosing the shortest route to the exterior, one she had discovered and memorized from her previous journeys. She soon widened the exit, got out, moved down the south side of the nest into the ring of attacking robber ants, and was set on by four of them and quickly killed.

Nest A was now queenless. A feature of ant life is the difficulty of communication. The wood ants would not know for some time that they had no queen, because such knowledge is conveyed by the lack of the queen substances in the food circulating around the nest. The message would take one or two days to become generally known. Under the prevailing battle conditions this was advantageous to the wood ants, since they did not know the bad news and continued to fight with the utmost ferocity. In the past, in human wars the loss of the commander has often led to the defeat of that side. For instance, a stray arrow killed King Harold at the battle of Hastings in A.D. 1066 and hastened the Norman Conquest. Had the English soldiers not known of his death, the result might have been different.

The wood ants, unaware of their loss, finally drove the robber ants away, with the loss of only about a hundred pupae and some five hundred workers. The attackers suddenly gave up as the sun sank and retired to their nest with the little booty they had obtained, most of which they eventually ate. The robber-ant nest was large, so its inhabitants were more anxious to get food than to increase their supply of workers, whether slaves or *sanguinea,* a course of action brought about by the low protein content of the food circulating among the ants. The need for food overrode the powerful urge to care for the young.

8

SEPTEMBER

IT was not until the early days of September that the ants in Nest A realized they had no queen. After the robber-ant attack had been driven off, there was a great deal of activity devoted to putting the nest to rights again. All the eggs, larvae, and pupae that had been carried to safe depths or piled on top of the nest ready for evacuation had to be returned to the brood chambers and galleries, and suitable spots had to be found for the pupation of the mature larvae. When Labora and her fellow attendants on the queen returned to the royal apartments, they could not find their mistress. This did not particularly distress them, because the chambers and galleries were so impregnated with the queen's exudations that no acute loss was felt. They did look for her, as if aware of her desire to escape, but in moving around on this quest they were repeatedly distracted by the urgent work going on. For instance, Labora, while traveling over the top of the heap trying to trace Regina I, came across bands of nurse ants moving pupae back into the nest and at once started to help them, ceasing only when all had been moved to safety. Ants' attachment to pupae is phenomenal. Similarly, Labora and her colleagues helped in the clearing and rearrangement of the area that had been occupied by the robber ants. The intruders left behind were all cut up, the soft parts being used as food and the chitin disposed of as rubbish.

Food was passing around the colony in the usual way, but, as there was no queen, the antifemale pheromone in it was gradually dropping, and naturally the first group of ants to notice this was the group of royal attendants, including Labora.

It is obviously fatal for an ant colony if the queen lays no eggs. The situation can be remedied by the nest's taking in a new queen at the time of the sexual swarming, but, as the flight had already taken place in September, it was too late.

A Trojan war could be the solution: stealing a queen from another wood-ant nest. By September 8 Labora and the former queen's attendants were becoming anxious over the absence of the quietening and satisfying queen chemicals in the circulating food. Foragers from Nest A, ever anxious to increase supplies, had trespassed into Nest G territories and numerous fights had taken place; ants on both sides were killed. Dead ants are usually eaten by the victors (a point made by an old-fashioned ant in Samuel Spewack's play is that to be eaten by enemy ants was an expected and honorable death). Nest A ants carried some of the dead Nest G workers back to their home. Labora and her colleagues noted that these ants with their alien smell did, however, have queen substances in them, and they set out to find their source. They soon came across the disputed territory and made a difficult reconnaissance toward Nest G, fighting off several attacks but finally placating a sentry with an offer of food. Labora managed to enter some of the surface galleries and checked the presence of queens in the nest by the exudations on the gallery walls. The desire for the queen substance and its discovery in Nest G led almost automatically to the organization of a raid. Labora and her companions returned to their own nest and started an agitation that lasted all night. They displayed bits of ants captured on Nest G, tapped antennae repeatedly, carried their companions out along the road they should take in the attack, rubbed faces with their legs, and generally encouraged the martial spirit. Soon after daybreak a column of Nest A wood ants was advancing on Nest G.

Their method of making war was quite different from that of the robber ants. They used no flanking movements or elements of surprise but attacked in the early morning in a solid, purposeful column about a foot wide, trying to penetrate the enemy area by sheer weight of numbers. The urgency of their need can be seen from the desperate measures they were taking. Numbers were what counted (God, Voltaire says, is on the side of the big battalions). Nest A was at a disadvantage in attacking its neighbors in G, for G, being a multiqueen nest, was much bigger. Naturally, the attack was resisted. The Nest G workers poured out of the many exits as the alarm chemicals circulated around the nest, and countless sanguinary battles took place. The A's managed to enter a few surface galleries but never lasted long in them. They never got near a royal chamber, nor did they capture any pupae. Their column was halted, pushed back, and ignominiously routed. They left three thousand dead on the field, and Labora was lucky to escape with her life. As the initiator she was largely behind the lines, recruiting and organizing the attack—the generalissimo, in fact.

The philosopher may compare the ants' struggle with one of our own civil wars. Had a reasoned discussion between the rivals been possible, the battle need never have taken place. Nest G, after a prosperous season, had surplus princesses and even had to kill some of those attempting to establish themselves in the colony. They could easily have spared some of their surplus. The ants, having no language, could not discuss the matter or make such an offer, but man, without this limitation, should not have to suffer from such intraspecific rivalry.

Nest A had thus suffered heavy losses and obviously could not solve its problem by fighting for a new queen. It had to either perish or find some other remedy. In the absence of a queen some workers can start to lay eggs, which in general obey Dzierzon's Law. Dzierzon, a Polish scientist, found out in the mid-nineteenth century that unfertilized insect eggs give rise to males and fertilized ones to females, the females becoming queens or workers according to how the larvae are fed, as noted before (see page 51). Since the workers carry no sperm, only males could be produced; but, as the male ants do no work in the colony, their production late in the season is not of much use. Had the queen been lost in May or June, a new generation of males could have been produced and a new queen taken in after the flight of the sexuals. With no workers being bred, things did not look very hopeful for Nest A. If it could not get a queen until the next swarming, in July, the ants would have to live on the existing work force until then or make a raid on another wood-ant nest and capture a queen—both risky alternatives. Fortunately for them, another course of action presented itself.

At this point, it is as well to examine the unusual sexual reproductive system of ants. It is a method found only in the social insects and one quite different from that prevailing in other animals. The subject has been admirably set out by Richard Dawkins in *The Selfish Gene.*

Colonies of social Hymenoptera are divided by Dawkins into bearers and carers. The bearers fertilize and lay the eggs, and the sterile carers tend the young and bring them to adulthood. There can thus be a conflict of interests between the genes of these two groups, and some consideration must now be given to the gene struggle and the gene content of males, queens, and workers.

In general, a worker ant leaves no offspring; therefore, the death of a worker has no direct effect on the colony's future gene content. However, such a death can have a slight indirect effect, because it has reduced the work force supplying the bearers with food and other attentions. The workers—the carers—can also influence the future composition of the colony by the kind of care they give to the brood; for instance, they control the number of princesses produced.

In most animals, when conception takes place the new individual—the zygote—derives half its chromosomes, carrying the genes, from the father and half from the mother. Consequently, each non-Hymenopteran spermatozoon of the father is different, as is each egg of the mother, and the resulting offspring can differ from their parents. Close similarities to parents arise largely from the fact that genes tend to move in groups. The differences from parents shown by offspring are the basis of natural selection, and those gene combinations best fitted to survive under the ruling conditions are the ones that, in fact, do survive. But in ants the males arise from unfertilized eggs: the males have no fathers. Consequently, each cell of each male has only one set of chromosomes, bearing genes derived entirely from his mother. Thus, male sperm are all the same and not all different, as in other animals.

The females have the usual two sets of chromosomes, one derived from the father and one from the mother, but as the fathers' sets are all alike as far as the father is concerned, the offspring are identical twins. A female ant (queen or worker) is thus more closely related to her sisters than to her own children of either sex.

The struggle for existence is between the genes, and thus it is more advantageous for the genes in a female worker ant to predispose her mother to turn out sisters rather than children, which is probably how the sterility of workers evolved. In this way the scarce resources—food, labor, nest space, and so on—are used to best advantage for those particular genes. The workers must, however, induce the sister-making machine—their mother—to do just that: to turn out more females than males. R. L. Trivers and H. Hare calculated that the two optimum sex ratios to serve the ants' interests are for a mother the usual 1:1 but for a sister 3 females to 1 male (R. Dawkins). In other words, the queens "want" a 1:1 ratio and the workers a 3:1 proportion. In investigative work Trivers and Hare found that in nineteen species of ants, victory in the conflict of interests usually went to the workers. The sex ratio of females to males in ants tended to be 3:1, account being taken of the size of the different insects. Obviously, the biomass, or weight, of the different groups is the measure of the resources put into that group. The workers control the nurseries and brood and thus are more influential than the queens, which just lay eggs.

In a nest containing slaves in any number, different conditions arise. The slave workers are unaware that they are slaves and, in attending their mistresses' brood, behave as if they—the slaves—were caring for the young of their own species in their distant home nest. In the case of wood ants (*rufa*) enslaved by *sanguinea*, the genes of the slave *rufa* nurses thus induce a course of behavior favorable to the production of *rufa* offspring, activities that will be different from the line of conduct best suited to the production of *sanguinea*. This gives the slaver queen (*sanguinea*) an opportunity of altering the sex ratio toward one favoring her own preferred proportion of 1:1, because changes made by such queens are not likely to be spotted by *rufa* workers. *Rufa*s take no countermeasures, because the *sanguinea* brood is completely alien to them. They have no genetic knowledge of what is the best thing to do for *sanguinea*. Trivers and Hare examined the sex ratio in two slave-making species (*Harpagoxenus sublaeris* and *Leptothorax duloticus*) and found that the sex ratio in the brood was 1:1, showing that in slave-owning nests the queens held the power.

Let us now return to Labora. She was an active ant and was among the first to notice the lack of queen substances in the food. In the absence of the inhibiting hormone, her ovaries began to develop and she started to lay eggs. Her former colleagues carried them to the egg piles, gave her food, cleaned her, and in due course treated her like a queen, feeding her largely on labial secretions. The first eggs she laid hatched to males, which not only were of no use to the colony in winter, spring, and early summer but made demands on the autumnal food supplies as well, lowering the amount available for establishing food reserves in the ants' bodies, for there was no foraging during winter.

The biology of Nest A (and of wood ants in general) was such that Dzierzon's Law could to some extent be overcome, an important factor in the survival of the colony. As it turned out, Labora was producing two types of eggs: some about one twenty-fifth of an inch in diameter, the others about one-fortieth of an inch. The big eggs produced males, but the small ones were to prove the salvation of Nest A. She was able to retain some of these eggs in her oviduct until chromosome doubling took place; it was equational, as were later divisions, and the result was the production of diploid eggs that would produce females.

Equational division means that in the sexual cell the chromosomes start dividing longitudinally and result in two equal segments being incorporated in the daughter nuclei. The nucleus is thus diploid; it has two exactly similar sets of chromosomes. These sets, however, have not been obtained one from each haploid parent, as would be the case in normal sexual reproduction, but both from the mother. As a result, they are of the same genetic makeup as the mother. The grubs from Labora's eggs could then be raised as queens or workers. Labora continued to lay eggs, mostly small ones, until halfway through the month, though at nothing like the rate of a true queen. The additional workers, hatched toward the end of the month, to which she gave rise, enabled the colony to pass the winter successfully and to be in a good state to take advantage of conditions in the following spring and summer. Labora herself started to secrete the antiqueen substance, which prevented other worker ants from laying eggs. She was thus supreme in the nest.

The nurse ants ate most of the male grubs produced.

The production of females in defiance of Dzierzon's Law is a strange thing. Of course, no sperm are involved. It is the equivalent of the parthenogenetic production of aphids. Labora's female offspring were like twin sisters and of exactly the same gene content. In this respect they were just like cuttings from a plant. As Labora's genetic composition was that of an activator ant, she certainly improved the status of Nest A, as all her offspring were activators, too. Of course, a nest can have too many exciters or leaders. All chiefs and no Indians does not make for efficiency.

Apart from *Xenodusa*, from which Nest A had had a narrow escape, there were other troubles. For instance, nematode worms (*Mermis* spp.) attacked a few of the workers, growing inside their abdomens until they were swollen to double the size and discharging a mass of eggs when the unfortunate ant died. The worm eggs lay around in the corridors ready to infect passing ants.

Another danger was an attack by fungi, which warmth and high humidity within the nest encourage. The ants did secrete a fungicide, but it was not effective against all species and strains. Occasionally a few ants would be attacked by a white mold. The infected ant usually left the nest, climbed a grass stalk, hung on with its forelegs, and died, the fungus sporulating on the dead insect.

9

OCTOBER

A few of Labora's worker offspring were emerging from their pupal cases at the beginning of October, adding to the labor force so much depleted by the senseless attack on Nest G. There is no doubt that Labora's diploid eggs had saved the colony. Some foraging was still being done, but the ants saw the sun's zenith sinking every day in the sky and felt the temperature dropping night by night; they knew that preparations must be made for hibernation. The nurse ants still paid attention to the pupae, moving them to the warmest spots during the day and deep down into the nest at night. October 15 was very cold and wet, so the pupae were left in the underground chamber that day. On October 17 some of the older nurses ventured outside the nest, familiarizing themselves with the Nest A tracks and aphid trees. The information they obtained would be of immense value the following spring, because they would be able to go straight to "their" trees and establish their claim to them before rival ants from, say, Nest G took over the area. Territory in the ant world is often acquired by being the first arrival at a desirable spot and mounting a holding garrison there.

On the same day the nurses also brought up the pupae to the sun-warmed side of the nest, and a few of them hatched. Those that did not were carried back at night and saw the daylight no more. Toward the end of the month most of them were eaten. The labial glands of Labora's attendants decreased in activity; by October 14 no more royal food (labial secretion) was fed to the substitute queen. She laid her last egg of the year on October 20. An attendant carried it to an egg pile in one of the upper galleries, but the weather was too cold for development and by the end of the month the whole egg pile had been eaten. Labora had prepared for winter by building up a considerable fat reserve in her body. She and her household now went down into one of the deep chambers in the nest. More and more workers were attracted to their second queen, and the slowly moving winter ball of ants formed. It stayed in this gently turning mass until the following spring.

Labora passed the winter successfully. The nest resumed activity the following April, and Labora started laying both large and small eggs again, the large ones producing males. Quite a sizable male swarm from Nest A took to the air in early July and attracted the attention of the young princesses from Nest G and other wood-ant colonies in the area. Many matings took place and once again the incipient *fusca* nests were in great demand. About this time Labora, the deputy queen who had saved the colony, was growing weaker. She could no longer produce the same large quantity of attractants that secured the attention of her staff, who in return let their supplies of labial secretion fall. As a consequence, Labora laid fewer eggs. Was the nest again in danger of extinction?

Another effect was that Labora could no longer produce so much of the antiqueen substance, resulting in the Nest A workers being less aggressive toward the young, recently mated females now eagerly searching for homes. In the middle of July one of these (Regina XX) placated a guard, entered the nest, exchanged food and queen attractant with a few foragers in the galleries, and hid for twenty-four hours, during which time she absorbed the nest smell. Regina XX then established herself in an empty chamber toward the top of the nest. She soon attracted a body of attendants and began laying eggs.

There were only two solutions to this situation: either the colony would become a multiqueen nest or the queens would fight until one was victorious.

Labora, though deputy queen, had not entirely lost her worker instincts. One day toward the end of July she felt an urge to go outside and forage. Her track happened to lead her along galleries passing Regina XX's quarters. The two queens were but an inch apart and a contest, in the pitch dark, seemed inevitable. The usurping queen was supported by attendants seduced from Labora's court by greater rewards (more queen-attractant substances), so that Labora was hustled around by them. They did not attack her directly, but pushed her away from the new queen's area. It was as if they hesitated to commit the crime of regicide, though the fact was that Labora still had enough of the queen smell about her to inhibit direct attack from the workers. Sensing that something was afoot from the agitation of her workers, Regina XX moved out from her chamber into the gallery Labora was being urged to vacate. Regina XX was, naturally, not restrained by a chemical identical to the one she herself was producing. In fact, it had the opposite effect: she immediately recognized a rival, one apparently attempting to enter her territory. A willingness to share a nest with a number of queens, or to have passages running to daughter nests, is a characteristic of the species, but Labora's presence seemed like a direct assault from an upstart worker, since Labora's comparatively small size identified her as of worker origin. A real queen ant, of course, would not "despise" a worker queen, and it is being painfully anthropomorphic to suggest it, but Regina XX would know that she had to defend her territory. Her genes had preconditioned her to defend their interests; so she attacked.

The young queen knew she had the advantage of being bigger than Labora, but, from the point of view of gene propagation, she had other advantages as well, of which she was unaware. She was better fitted for producing eggs and had the great advantage (as far as the survival and extension of the species went) that she could lay eggs with a varied genetic makeup. Her eggs could produce a range of ants with differing qualities. Among them might be a line leading eventually to a super ant (for instance, one able to communicate as well as a bee), whereas Labora's eggs could only reproduce more Laboras—identical twins, so to speak. The production of a super ant was unlikely, but small advantages in the offspring could be accumulated by the sexual method.

Sex is just a system of mixing genes. It is a method enabling the species to survive by means of certain automatically selected individuals, or even small changes in their own habits, or of alterations in the environment. Labora's offspring, on the other hand, good as she was, could do only what Labora did and if, over the years, the whole nest were to be composed of Labora's descendants, then, if conditions turned against her qualities, the nest risked extermination. In other words, a colony of mixed race is more likely to have some individuals survive a catastrophe than a homologous colony of identical individuals. This is not to say that alterations, brought about by mutation of the genes, cannot occur in asexual reproduction, but changes induced by the sexual method are generally more usual and quicker, too.

The agitation among her staff and the emission of a small amount of alarm substance led Regina XX to leave her home. She smelled her way toward Labora and met her head-on in a narrow and smooth passage up which Labora had been climbing. Regina XX's weight and her forward rush downward gave her an initial momentum. She pushed Labora back down to another horizontal and wider gallery, and the two ants fell separately a tiny distance apart. Labora started to retreat toward her own quarters, but Regina XX was by now roused to secure a final answer. Her antennae quickly smelled Labora's track, and she hurried after her. Labora now took another branch passage downward, but unfortunately one of her own attendants was coming up it, forcing Labora to retreat and giving Regina XX an opportunity to seize one of Labora's hind legs as she backed out of the passage. The offending attendant came out and palped the two struggling ants, but, becoming confused because they both had the correct smell of Nest A, she wandered away after Regina had scratched her abdomen with her mandibles. Regina XX followed up her advantage over Labora in the usual way. She stretched the leg she had secured, twisted her abdomen around, and, locating the correct direction with her antennae, sprayed the tarsal joint with formic acid, incapacitating it. After much more struggling, in which Labora, too, ejected the poison spray but with less directional accuracy, Regina XX managed to seize Labora's left antenna just above the scape and bite it off. In the dark this was a serious loss, because it destroyed much of Labora's sense of smell, in particular its stereoscopic power. As mentioned, ants smell with their antennae, having in effect two very motile noses enabling them to sense the direction of a smell with great accuracy, a matter of much importance in their underground life. With the use of two waving antennae, a pair of palps and two forelegs, ants in the dark have no need of sight to conduct their daily affairs.

With the loss of half a nostril Labora was slower in finding out just where her opponent was. The end was now inevitable. Although Labora did manage to spray Regina's legs with acid and temporarily immobilize two of them, the new queen was mistress of the situation. She caught her rival's good antenna in her jaws and amputated it at the base, leaving Labora pretty well helpless in the dark. Regina then ended the contest by cutting off Labora's head.

Regina XX now became supreme in the nest. At first Labora's immediate staff were at a loss; then some of them carried the body down to the sorters (nothing must be wasted in the ant economy). Others went foraging, and some even joined the new royal household. As in the world of humans:

> *Treason doth never prosper: what's the reason?*
> *Why, if it prosper, none dare call it treason.*
>
> (Sir John Harington, 1561–1612)

Once again, Nest A had been saved. Harsh as it might seem, the solution imposed was a good one for the survival of the genes conditioning behavior in Nest A.

At first sight, ants seem to live in a society motivated by emotions similar to our own, but the resemblance is really only superficial, as the account of a year in the life of Nest A shows. The similarities (such as the battles between rival nests, or nations, of the same or different species) are due to the fact that any two societies, by the very fact that they are societies, must show similarities. It is like saying any two even numbers show similarities—they are both divisible by two, for instance. A society is "a system or mode of life adopted by a body of individuals for the purpose of harmonious co-existence or for mutual benefit, defence, etc.," as defined in *The Oxford English Dictionary*.

The human and ant societies are similar because they are societies, but their motivations and working methods are very different. As Darwin has pointed out, societies have arisen through the struggle for existence. He wrote, "In social animals natural selection will adapt the structure of each individual to the benefit of the community."

The term *structure* refers to both the actual physiology of the animal and to its mode of behavior, or instinct. In Darwin's words, "It will be universally admitted that instincts are as important as corporeal structures for the welfare of each species, under its present conditions of life." Darwin noted the great difference existing between human and insect societies and regarded the latter more as families.

Not that other, earlier, naturalists were unaware of the importance of these matters to social animals. For instance, the eighteenth-century American author and naturalist Samuel Williams, in his account of Vermont, was intrigued by the social life of the beaver. He noticed social behavior in beavers that was the same as that subsequently found in ants. In his words: "When the beaver is found in a solitary state he appears to be a timid, inactive, and stupid animal. . . . When combined in society, his disposition, and powers assume their natural direction, and are excited to the highest advantage: Everything is then undertaken, which the beaver is capable of performing (dams, fishing etc.). . . . The society of beavers seems to be regulated and governed, altogether by natural dispositions, and laws."

Williams also noted that (again, like ants) beavers had no chief or leader, presented every appearance of a perfect democracy, and showed a principle of perfect equality and "the strongest mutual attachment." It was the young republic speaking.

The selection of favorable behavior patterns can be seen to be particularly important in the case of wood ants. For instance, their use of *fusca* nests saves them a lot of effort in starting new colonies. Their arrangement with the aphids (protection against enemies in return for honeydew) is a commercial transaction satisfactory to both parties. That the ants have survived for such a vast length of time is due more to the gradual accumulation of genes inducing favorable behavior controlled by chemicals circulating in the nest than to modifications of their bodies, which have been very slight over the millennia.

It is the practice of societies to produce goods in common for the benefit of the whole, and societies, such as those of ant and man, have found that specialization by sections is the easiest way to produce such goods and services. For example, the nest queen lays the eggs, nurses tend the young, foragers gather food, and sorters sort. The butcher, the baker, and candlestick maker follow (or used to) their respective occupations in our society, and, in general, each person is providing services or goods in excess of his or her own requirements. In both societies these surplus products, or profits, are distributed to the community in different ways. The ants effect this distribution by what seem to us somewhat primitive and wasteful means. The ants' methods are constrained by the physical and mental facilities at their disposal; in particular, communication difficulties impose certain inefficient practices on them, such as foragers laboriously bringing in objects that are immediately ejected from the nest. Humans distribute their surplus product by a variety of differing political and social structures that are much more efficient in terms of benefit to the species but not necessarily to the ecology of the world as a whole.

In terms of the gene struggle, ants do not have it all their own way. Wood ants waste a large amount of their surplus product, a wastage discernible as being on two levels. First, there is the raising and feeding of a large number of unneeded males and females, and second, there is the provision of food and care for a crowd of guests in the nest, some of them positively dangerous. In the first case, the wastage may be more apparent than real. From the point of view of gene survival, the production of large numbers of apparently unwanted sexuals is probably the best method ants have, in view of their limited physical and mental attributes. A large swarm of sexuals ensures that at least some mated females survive to continue the species. The fact that the system exists and that ants are a successful order suggests that the wasteful method is about the best one for them, Panglossian though this observation may be. If not the best of all possible worlds for them, it is close to it.

The second case, the feeding of guests and predators, can be seen as a partial victory for some rival genes, those in some of the guest animals. For instance, the *Xenodusa* genes have been able to implant such deceptive powers in their "survival machines" as to induce ants to waste their substance in feeding and supporting their enemies.

William M. Wheeler, impressed with this curious state of affairs, could as a result be accused of today's worst crime among biologists—anthropomorphism. He contrasted ant and human behavior toward guests and, making a very effective point, wrote: "Were we to behave in an analogous manner we should live in a truly Alice-in-Wonderland-Society. We should delight in keeping porcupines, alligators, lobsters, etc., in our houses, insist on their sitting down to table with us and feed them so solicitously with spoon victuals that our children would either perish of neglect or grow up as hopeless rachitics."

Over a wider field, the ants' surplus product, the unneeded sexuals, do form food for other animals, particularly the pileated woodpecker (*Ceophloeus pileatus*) and the golden-shafted flicker (*Colaptes auratus*).

On the other hand, ants do not waste the dead killed in wars but eat them, whereas we either burn them or bury them too deep to be of much value to the land as manure.

The economics of food production by ants compare quite favorably in some respects with those of man. E. O. Wilson estimated that 65,000 wood ants, weighing 700 grams, made 300,000 foraging trips a day and collected at least 800 grams of food, of which 44 percent was honeydew. That is, one ant, weighing 11 milligrams, produced just over 12 milligrams of food a day (a little more than its own weight). In 150 days of foraging it would collect about 1,850 milligrams of food, or about 168 times its own weight per summer.

In a field of wheat a man, weighing about 150 pounds, equipped with modern machinery and working for, say, two days per acre per season will produce about 3,500 pounds of grain, or only 23 times his own weight. But he leaves the ant behind when time is taken into consideration. The ant secures its own weight of food per day, using no tools. The man, increased to two men to allow for the manufacture of the plow, drill, harvester, and fuel to run them, produces food at the rate of just under six times his weight per day:

$$3{,}500 \div (2 \times 2 \times 150) = 5.83$$

It is interesting that, left to his own devices and even allowing him a few primitive tools, a man is a much less efficient producer than an ant. He succeeds only because he has used his surplus product more efficiently (in inventing tools and political systems), which is brought about primarily by his powers of communication.

Ants, as seen in this story, are poor communicators compared with man—no speech, no writing, no bee dance. Speech is a great gift to us, but the ants manage without it. Speech, said Talleyrand, in a cynical mood, "was given us to conceal our thoughts," and certainly the ants' thoughts are pretty open to all the community, being largely expressed by the emission of chemicals into the surrounding air; that is, by smells, or by incorporating pheromones in the food.

Ants also use touch and gesture to communicate. Gesturing is not of much use to them in the dark, but it is of value in daylight. It can be of importance in communication in many forms of life. It can even replace speech in man. For instance, the Cistercian monks held speech to be sinful, except in religious exercises, and invented a gesture language to replace it. Leibnitz prepared Latin and German dictionaries of their system. In passing, it may be of interest to mention some of these gestures, many of them being obvious, others strange. They illustrate how one sense can replace another and give a clue as to how ants do without certain senses we find almost essential.

Two gestures nearly everyone uses are shaking and nodding the head for no and yes. Stranger movements are touching the little finger to the tip of the nose, meaning "fool," and the forefinger there, signifying "wise man." Two fingers on the right side of the nose were for "friend," on the left "enemy," and so on.

Lucian (A.D. 37–65) recorded that a barbarian prince at Emperor Nero's court saw a miming actor perform so well that he understood his song. When Nero asked the prince what gift he would like, he requested the actor. It was difficult, said the prince, to get interpreters to speak to his many subject tribes, as there were so many different languages, but the actor would answer the purpose perfectly. History does not record if the unfortunate actor was banished among the barbarians—the penalty of success. Should our modern actors be put to watch ants? So many of them now cannot be heard, so a gesture language could be useful! At present, we know only the meaning of one ant movement with any certainty: excitedly running about—kinopsis—means that something important is afoot.

In 1810 Huber suggested that ants used a tapping code with their antennae, a theory much favored since by some authors of popular works. Three taps meant this, five taps that, and so on. Modern myrmecologists, such as R. Chauvin and J. Sudd, very much doubt if this is so. It is now thought that the antennal blows just mean "something's happening, follow me," after which the following ant imitates the action of the activator and the task is eventually completed. Taste and smell, as we have seen, are most important to wood ants. The whole colony is run by it. Human societies are run by thought processes; mistaken though they may be at times, they secure a better life—one with less suffering, more leisure, and a greater understanding of the universe—than do the chemically controlled societies of ants. Of course, the suggestion has now arisen from the behaviorists, among whom B. F. Skinner is an outstanding example, that man is just as much a slave of his genes and environment as are ants. Free will and the autonomous man have been abolished, says Skinner.

If what we do is entirely conditioned by our heredity and surroundings, and objective reasoning is of no importance or even does not exist, then it is surprising that up to the present we have not been in much greater competition with ants. A few clashes do take place. The fire ant is an annoyance. Leaf cutters and Argentine ants can be crop pests, but the damage done to us is not great. However, our existence on earth has been short compared with that of the ants; perhaps there has not yet been time for a serious conflict. If the behaviorists are right, we and the ants (we are both omnivores) will eventually be in competition for food as our respective populations grow, and it could be argued that the few pest ants we have are the start of this rivalry. If the rivalry grows, eventually we shall have to fight it out with the ants. But the behaviorists' suggestion does not seem likely to be true. Humans every day, everywhere, are obviously, consciously, making choices irrespective of their genes and surroundings. And the choices are usually motivated with beneficial objectives in view. For instance, every day more people decide to go to work than to rob a bank. However, if we do have to fight the ants, we shall win, for we are more clever. We have defeated many insect plagues—locusts and cattle ticks, for example. The ants will take over only if we first destroy ourselves.

APPENDICES

NOTES

1. "Go to the ant, thou sluggard: consider her ways and be wise: which having no guide, overseer or ruler, provideth her meat in the summer, and gathereth her food in the harvest." Proverbs VI:6–8

2. Aliphatic compounds have a chain of molecules, as distinct from ring-arranged molecules.

3. Not that a head is essential to an ant. C. Janet observed a *rufa* that lived for twenty-nine days without a head!

4. Soma—the drug with no damaging aftereffects, which kept the masses contented in Aldous Huxley's *Brave New World.*

5. S. T. Coleridge: "For he on honeydew hath fed and drunk the milk of paradise."

6. Attempts to get rid of ants by offering them poisoned syrup are only successful if the food is but lightly doctored. Heavily poisoned syrup is rejected by the sorters if it does not actually kill the forager, and after a time the ants leave it alone.

7. In my view, cocaine was a factor in the destruction of the Inca empire. Under the Incas, cocaine was strictly controlled and could be taken only by aristocrats and the relay runners (the *chasquis*). After Pizarro's Conquest, the previously well-guarded stores were left unprotected and the coca leaf became widely distributed, inducing addiction and apathy in the population. Any will to resist the Spaniards collapsed. In fairness it must be stated that the Spaniards had no idea of the nature of the coca leaf.

BIBLIOGRAPHY

BOOKS

Chauvin, Rémy. *Animal Societies: From the Bee to the Gorilla.* Translated by G. Ordish. London: Gollancz, 1968.

———. *The World of Ants.* Translated by G. Ordish. London: Gollancz, 1970.

Compton, John. *Ways of the Ant.* London: Collins Sons, 1953.

Darwin, Charles R. *Origin of Species by Means of Natural Selection.* London: John Murray, 1860.

Dawkins, Richard. *The Selfish Gene.* Oxford: Oxford University Press, 1976.

Dethier, V. G. *The Physiology of Insect Senses.* London: Methuen, 1964.

Donisthorpe, Horace St. J. K. *British Ants: Their Life-History and Classification.* London: Routledge, 1927.

Fabre, Jean-Henri. *Social Life in the Insect World.* 4th ed. Translated by B. Miall. London: Fisher Unwin, 1916.

Ferenczy, Arpad. *The Ants of Timothy Thummel.* London: Cape, 1924.

Harington, Sir John. *Epigrams.* Book IV, no. 5. London: J. Budge, 1618.

Huber, P. *Vie et moeurs des fourmis indigènes.* Paris, 1810.

Huxley, Aldous. *Brave New World.* London: Chatto & Windus, 1932.

Hyams, Edward. *Morrow's Ants.* London: Lane, 1975.

Leibnitz, Gottfried Wilhelm von. *Opera omnia.* IV. Geneva: Dutens, 1768.

Lovelace, Richard. *The Poems of Richard Lovelace*, vol. 2. Edited by C. H. Wilkinson. Oxford: Clarendon Press, 1925.

Millar, George H. *A New, Complete, and Universal Body, or System of Natural History.* London: Alex. Hogg, n.d. 1785 (?).

Muesebeck, C. F. W.; Krombien, K. V.; and Townes, H. K. *Hymenoptera of America North of Mexico.* Washington, D.C.: U.S. Department of Agriculture, 1951.

Oxford English Dictionary, The. Oxford: Clarendon Press, 1971.

Pope, Alexander. *The Poetical Works*, vol. 2. Glasgow: Foulis, 1785.

Skinner, B. F. *About Behaviourism.* London: Cape, 1974.

Spewack, Samuel. *Under the Sycamore Tree.* New York: Dramatists' Play Service, 1960.

Sudd, John H. *An Introduction to the Behaviour of Ants.* London: Edward Arnold, 1970.

Voltaire. *Letter to M. le Riche.* February 6, 1770.

Wheeler, William Morton. *Ants: Their Structure and Development.* New York: Columbia University Press, 1925.

Williams, Samuel. *The Natural and Civil History of Vermont*, vol. 1. 1809; 1794.

Wilson, Edward O. *The Insect Societies.* Cambridge, Mass.: Harvard University Press, Belknap Press, 1971.

———. *Sociobiology: The New Synthesis.* Cambridge, Mass.: Harvard University Press, Belknap Press, 1975.

ARTICLES

Auclair, J. L. "Aphid Feeding and Nutrition." *An. Rev. Entomology*, vol. 8, 1963.

Jander, R. "Die optische Richtungsorientierung der roten Waldamesie (*Formica rufa* L.)." *Zeitschrift für vergleichende Physiologie*, 40. Berlin, 1957.

Kloft, W. "Recognition of Aphids by Ants." *Biol. Zentr.*, 54. Berlin, 1959.

Löfqvist, Jan. "Formic Acid and Saturated Hydrocarbons as Alarm Pheromones for the Ant *Formica rufa.*" *Journal of Insect Physiology*, 1976.

Regnier, F. E., and Wilson, E. O. "The Alarm-Defence System of the Ant *Acanthomyops claviger.*" *Journal of Insect Physiology*, vol. 14, 1968.

Trivers, R. L., and Hare, H. "Haplodiploidy and the Evolution of the Social Insects." *Science*, vol. 191.

Watt, W. W. "Had Enough? Vote Ant." *The New Yorker*, November 1946.

Way, M. J. "Mutualism Between Ants and Honeydew-Producing Homoptera." *An. Rev. Entomology*, vol. 8, 1963.

Wheeler, William Morton. "An Annotated List of the Ants of New Jersey." *Bulletin of American Museum of Natural History*, vol. 21. New York, 1905.

———. "Observations on Some European Ants." *Journal of New York Entomological Society*, vol. 17, 1909.

Wigglesworth, Sir Vincent. "Fifty Years of Insect Physiology." *XIIth International Congress of Entomology.* London, 1965.

Wilson, Edward O. "The Social Biology of Ants." *An. Rev. Entomology*, vol. 8. Palo Alto, Calif.: An. Reviews, Inc., 1963.

INDEX